Aqueous Acid–Base Equilibria and Titrations

Robert de Levie

Georgetown University & Bowdoin College

Series sponsor: **ZENECA**

ZENECA is a major international company active in four main areas of business: Pharmaceuticals, Agrochemicals and Seeds, Specialty Chemicals, and Biological Products.

ZENECA's skill and innovative ideas in organic chemistry and bioscience create products and services which improve the world's health, nutrition, environment, and quality of life.

ZENECA is committed to the support of education in chemistr

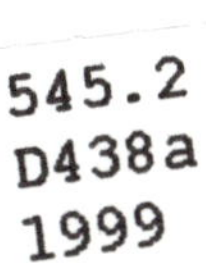

OXFORD
UNIVERSITY PRESS

OXFORD
UNIVERSITY PRESS

Great Clarendon Street, Oxford OX2 6DP
Oxford University Press is a department of the University of Oxford.
It furthers the University's objective of excellence in research, scholarship,
and education by publishing worldwide in

Oxford New York
Athens Auckland Bangkok Bogotá Buenos Aires Calcutta
Cape Town Chennai Dar es Salaam Delhi Florence Hong Kong Istanbul
Karachi Kuala Lumpur Madrid Melbourne Mexico City Mumbai
Nairobi Paris São Paolo Singapore Taipei Tokyo Toronto Warsaw

with associated companies in Berlin Ibadan

Published in the United States
by Oxford University Press Inc., New York

First published 1999

A catalogue record for this book is available from the British Library

Library of Congress Cataloging in Publication Data
(Data applied for)

ISBN 0 19 850617 1

Typeset by the author

Printed in Great Britain
on acid-free paper by
The Bath Press,

Series Editor's Foreword

Oxford Chemistry Primers are designed to provide clear and concise introductions to a wide range of topics that may be encountered by chemistry students as they progress from the freshman stage through to graduation. The Physical Chemistry series contains books easily recognised as relating to established fundamental core material that all chemists need to know, as well as books reflecting new directions and research trends in the subject, thereby anticipating (and perhaps encouraging) the evolution of modern undergraduate courses.

In this Physical Chemistry Primer Professor Robert de Levie presents an exceptionally clear and elegant introductory account of *Aqueous acid–base equilibria and titrations*. The book explains in precise terms the basic ideas and applications of a subject which is essential fundamental knowledge for all practising chemists. This Primer will be of interest to all students of chemistry and their mentors.

Richard G. Compson
Physical and Theoretical Chemistry Laboratory,
University of Oxford

Preface

This primer deals with a topic familiar to chemists, but perhaps in a non-familiar way. The traditional treatment of acid-base problems, and of titrations, uses a variety of approximations, but seldom provides the tools to evaluate under what conditions such approximations are valid, or even necessary. More importantly, the approximations are often inapplicable in the more complicated cases encountered in practice, leaving the student unprepared to handle those cases.

In this small volume we will start from the other end, by beginning with a complete graphical overview of the chemical equilibria involved. We will combine this with a simple but exact result, the *proton condition*, a conveniently condensed form of the mass and charge balance equations that is central to all kinds of pH problems. We then use diagrams to find appropriate approximations to solve the proton condition. This approach, pioneered by Niels Bjerrum (1914), was worked out in detail by Hägg (1940), and was propagated and further developed by Scheel (1955), Sillén (1959), and Butler (1964). It provides order-of-magnitude estimates of the concentrations of all species involved, so that one can make rational approximations, which in turn lead to more reliable answers. It provides approximate answers even in cases where the more traditional method fails completely.

Titrations dramatically illustrate the difference in approach. Traditionally, titrations are described in terms of piecemeal approximations, based on simple models that become cumbersome, increasingly inaccurate, or even inapplicable, when applied to polyprotic acids and bases, or to mixtures. Instead, we here start with exact expressions describing the entire progress of the titration. Such *exact* results are quite simple, and *remain* so as the systems involved become more complicated, thereby rendering approximations superfluous. The modern theory of titrations was developed by Butler (1964), Fleck (1966), Waser (1967), Willis (1981), and de Levie (1992b, 1996).

The problems discussed in this primer can all be considered as purely mathematical, and can be treated that way, but most chemists are uncomfortable with such a chemically sterile approach. Starting from approximations is not an acceptable solution either, because they are often of limited applicability and uncertain accuracy. In this primer I have attempted to find a middle way, by reducing the mathematical difficulty to a minimum without compromising either the underlying chemical complexity or the mathematical rigor. Graphs are used extensively to visualize the concentrations of all participants in acid-base equilibria, and as powerful aids in the computation, by illustrating the precise nature and severity of all approximations.

Almost all of the material covered in the first six chapters of this primer can also be found, in greater detail, in my *Principles of Quantitative Chemical Analysis* (McGraw-Hill, 1997). In this primer I have tried to reduce that material to its simplest form, in order to make it more readily available to a larger audience.

I am grateful to Mr Ming Wang and Prof W. Ronald Fawcett for their valuable suggestions. I hereby invite readers to give me the benefit of their comments and suggestions.

Chevy Chase, MD R. de L.
May 1999

Contents

1 Basic concepts

1.1 What are acids, bases, pH? And why bother?

Acids were already known in antiquity for their sour taste and for their power to solubilize metals. The Latin word for sour is *acidus*. In the early middle ages, the Arabs used nitric acid to separate silver from gold by selective dissolution. The use of vegetable dyes as acid–base indicators goes back at least to Robert Boyle (1627–1691). However, the nature of acids and bases was still obscure. In the eighteenth century it was believed that phlogiston, from the Greek word for flame, φλοξ (pronounced *phlox*), was the acidic principle.

Lavoisier observed that burning elements such as carbon, nitrogen, and sulfur in oxygen gave compounds that, when dissolved in water, produced acids. He therefore associated acidity with oxygen. Lavoisier concluded that oxygen was the generator of acidity, and therefore coined a name for it by combining the Greek word for acid, οξυσ (pronounced *oxus*), with the root of the word *gen*erator.

It was only after Davy showed in 1811 that hydrochloric acid contains no oxygen, and von Liebig introduced the concept of mobile, *replaceable* hydrogen in 1838, that acidity came to be associated with the presence of hydrogen rather than oxygen. Our present ideas about acidity were further developed by Arrhenius, who introduced the idea of *electrolytic dissociation* in 1887. The replaceable hydrogen then became a hydrogen *ion* which could dissociate from an acid, as in $HA \rightleftharpoons H^+ + A^-$. Likewise, a base was associated with the presence of a dissociable hydroxyl group: $BOH \rightleftharpoons B^+ + OH^-$. The Arrhenius definition is quite appropriate for aqueous solutions, because water itself can dissociate into H^+ (or, written in its hydrated form, as H_3O^+) and OH^-.

The concept of acid and base can be generalized in several ways. For example, in liquid ammonia, the NH_4^+ and NH_2^- ions play roles comparable to those of H_3O^+ and OH^- in water, so that NH_4Cl and $NaNH_2$ in ammonia can be considered as an acid and base respectively, just as HCl and NaOH are in water. This solvent-based classification of acids and bases is due to Franklin (1905).

In this book we will use an alternative definition given by Brönsted (1923), which emphasizes the complementary nature of acids and bases in aqueous solutions. It considers as an acid any substance that can donate a proton, and as a base any proton acceptor, i.e., the base in NaOH is OH^-. In this nomenclature, an acid that loses its proton becomes a base, and vice versa, so that one can consider conjugate acid–base pairs. This definition is independent of the nature of the solvent, and applies even in the absence of any solvent, as in the vapor phase reaction of HCl with NH_3 to yield NH_4Cl.

Examples of such conjugate monoprotic acid–base pairs are acetic acid and acetate, or NH_4^+ and NH_3. On the other hand, the definition can be context-dependent: bicarbonate can be considered a base when titrated with an acid, or an acid when titrated with a base.

At the same time, Lewis (1923) suggested a further generalization, by considering as acid any substance that can accept an electron pair, and as base any donor of such a pair of electrons. In water, this definition is equivalent to those of Arrhenius and Brönsted, since H^+ lacks electrons, while OH^- has a pair of electrons it can share. The Lewis definition has proven very useful in non-aqueous chemistry, such as in molten salts, but it has no obvious advantages in aqueous solutions, and will therefore not be used here.

The concept of pH (originally written as P_H) was introduced by Sørensen (1909) as the negative logarithm of the hydrogen concentration,

$$pH = -\log[H^+] \tag{1.1}$$

This logarithmic form is convenient to encompass the large range of values of $[H^+]$ encountered in aqueous solutions, from about 10 M to 10^{-17} M. Since relatively few solutions have proton concentrations above 1 M, the minus sign makes the pH a mostly positive quantity. Apart from the convenience of its logarithmically compressed scale, the concept of pH has stuck because the most common, electrometric method to measure acidity (i.e., with a glass electrode) yields a measurement approximately proportional to pH rather than to $[H^+]$. Here we will use the original, concentration-based definition of pH. Subsequent modifications of that definition have been more ambiguous, and will be discussed in the final chapter of this primer.

What is so special about pH? At this point you may well ask whether it is useful to devote an entire (though mercifully thin) book merely to the calculation, manipulation, and detection of pH. The answer is that pH is very important, not just in analytical chemistry, but in all of chemistry, and even in much of our daily experience.

More than three-quarters of the surface of the world is covered by water. More than three-quarters of our body, measured either in terms of weight or volume, consists of water. Apparently, life on this planet started in water, and water is still its central medium. Without water, plants cannot photosynthesize, agriculture would be impossible, and all life on earth would come to a halt. This gives the self-dissociation of water such a pivotal role.

Water is by far the least expensive and least polluting of solvents, and is therefore the preferred industrial solvent. Most industrial chemical processes in water require tight control of pH. Oxidation and reduction reactions often depend on pH as well, since electron transfer may be coupled to proton transfer. Metal corrosion also depends strongly on pH, so that even the use of water as an industrial coolant involves pH.

Life is a beautifully complex set of chemical reactions, finely geared to maintain and perpetuate itself. The central chemical agents of life are the enzymes, which catalyze the necessary reactions with exquisite selectivity.

Most enzymes are very sensitive to pH, and can function properly only in a rather narrow pH range. Even minute deviations from the norm may spell or foretell disaster, which is why blood pH is routinely measured to one-thousandth of a pH unit.

The acidity of aqueous solutions determines which metal ions are soluble in it, or are hydrolyzed and, perhaps, precipitate. Consequently, the availability of metal ions in the soil depends on its pH, which is why the soil pH largely determines which plants can grow in it.

The pH of rainwater is crucial for plant growth: acid rain is a major cause of deforestation in Eastern Europe, where unbridled industrialization without environmental controls belched enormous amounts of sulfur dioxide and nitrogen oxides into the air. Likewise, the pH of Athens rainwater may determine how long we can still enjoy the detailed features of the statuary on the Acropolis.

Acidity is one of only four solution properties (the others being sweetness, bitterness, and saltiness) for which we have specific taste receptors.

Even the pH of the paper on which this text is printed is important: acidic papers will deteriorate with age.

1.2 The mass action law

The fundamental law of chemical equilibrium is the *mass action law*, first formulated by Guldberg and Waage (1864), and further refined by them as well as by Horstmann (1873) and van 't Hoff (1877). Consider the equilibrium between the chemical species A, B, C, etc. on the one hand, and P, Q, R, etc. on the other,

$$aA + bB + cC + \ldots \rightleftharpoons pP + qQ + rR + \ldots \tag{1.2}$$

where $a, b, c, \ldots, p, q, r, \ldots$ are the associated *stoichiometric coefficients*. When we denote the corresponding concentrations by [A], [B], [C] etc., the mass action law then states that, at equilibrium, the ratio of the two products $[A]^a\,[B]^b\,[C]^c \ldots$ and $[P]^p\,[Q]^q\,[R]^r \ldots$ is constant. Consequently, an *equilibrium constant* can be defined as either

$$\frac{[A]^a[B]^b[C]^c\ldots}{[P]^p[Q]^q[R]^r\ldots} \quad \text{or} \quad \frac{[P]^p[Q]^q[R]^r\ldots}{[A]^a[B]^b[C]^c\ldots} \tag{1.3}$$

Which one of these two ratios is used to define an equilibrium constant is, ultimately, arbitrary, and is usually linked to the equally arbitrary way in which eqn (1.2) is written. Consequently it is different for different subdisciplines: in acid–base problems we use *dissociation* constants, while complexation equilibria are typically couched in terms of *formation* constants. The possibility of confusion can be minimized by specifying the *dimension* of the equilibrium constant used.

For a monoprotic acid HA the acid–base equilibrium $HA \rightleftharpoons H^+ + A^-$ is traditionally described through the *acid dissociation constant* K_a, defined as

$$K_a = \frac{[H^+][A^-]}{[HA]} \tag{1.4}$$

For the self-dissociation of water, $H_2O \rightleftharpoons H^+ + OH^-$, we similarly have

$$K_a = \frac{[H^+][OH^-]}{[H_2O]} \tag{1.5}$$

In dilute aqueous solutions, the term $[H_2O]$ is often nearly constant, in which case it can be included in the equilibrium constant, yielding an expression for the *ion product*,

$$K_w = [H_2O]K_a = [H^+][OH^-] \tag{1.6}$$

For bases we will use a similar formalism. For example, for the ammonium–ammonia equilibrium $NH_4^+ \rightleftharpoons H^+ + NH_3$ we write

$$K_a = \frac{[H^+][NH_3]}{[NH_4^+]} \tag{1.7}$$

where we treat NH_4^+ as an acid, and NH_3 as its conjugated base. Consequently it is unnecessary to use additional equilibrium constants, such as *base association constants* K_b.

The above formalism is readily extended to diprotic and polyprotic acids and bases, by using their *stepwise* dissociation constants. Thus, for a diprotic acid H_2A we have the successive equilibria

$$H_2A \rightleftharpoons H^+ + HA^- \tag{1.8}$$

$$HA^- \rightleftharpoons H^+ + A^{2-} \tag{1.9}$$

with the stepwise dissociation constants

$$K_{a1} = \frac{[H^+][HA^-]}{[H_2A]} \tag{1.10}$$

$$K_{a2} = \frac{[H^+][A^{2-}]}{[HA^-]} \tag{1.11}$$

Likewise, for a triprotic acid, we have

$$H_3A \rightleftharpoons H^+ + H_2A^- \tag{1.12}$$

$$H_2A^- \rightleftharpoons H^+ + HA^{2-} \tag{1.13}$$

$$HA^{2-} \rightleftharpoons H^+ + A^{3-} \tag{1.14}$$

with

$$K_{a1} = \frac{[H^+][H_2A^-]}{[H_3A]} \tag{1.15}$$

$$K_{a2} = \frac{[H^+][HA^{2-}]}{[H_2A^-]} \tag{1.16}$$

$$K_{a3} = \frac{[H^+][A^{3-}]}{[HA^{2-}]} \tag{1.17}$$

Again, the formalism is not restricted to acids, and applies equally well to diprotic bases such as ethylene diamine or sodium carbonate, to aminoacids, or to, e.g., ethylene diamine tetraacetic acid, which has four carboxylic acid groups and two amino groups. In all such cases K_{a1} applies to the dissociation of the first proton from the most highly protonated form, i.e., to the equilibrium $H_2CO_3 \rightleftharpoons H^+ + HCO_3^-$ in the case of Na_2CO_3, and to $H_4Y^{2+} \rightleftharpoons H^+ + H_3Y^+$ in the case of EDTA (where the fully deprotonated anion is traditionally given the symbol Y^{4-}).

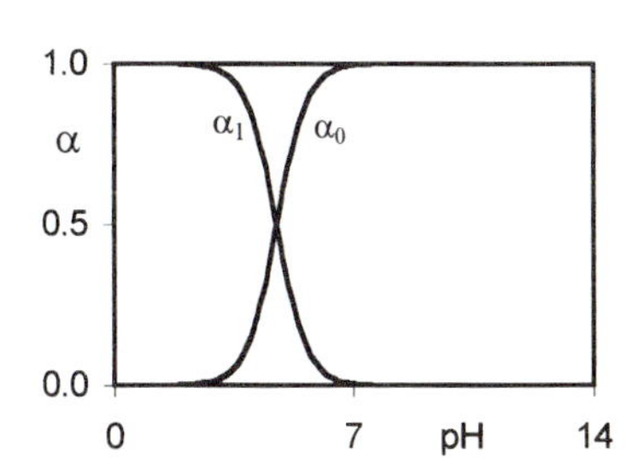

Fig. 1.1 The concentration fractions α_1 and α_0 (as a function of pH) of the acidic and basic forms, HAc and Ac^- respectively, of acetic acid, $pK_a = 4.76$.

1.3 Concentration fractions

We now define a few additional parameters. When we add a weak acid to water, part of the acid will dissociate, so that we will end up with a solution containing HA, A^-, H^+, and OH^-. We will normally have weighed the acid, and we will have used a volumetric flask in making the solution. In that case, we know the *total analytical concentration* of the acid, i.e., the sum of its ionized and protonated forms, which we will denote by the symbol C,

$$C = [HA] + [A^-] \tag{1.18}$$

We now define the concentration fractions α_{HA} and α_{A-} as

$$\alpha_{HA} = [HA] / C \quad \text{or} \quad [HA] = C\ \alpha_{HA} \tag{1.19}$$

$$\alpha_{A-} = [A^-] / C \quad \text{or} \quad [A^-] = C\ \alpha_{A-} \tag{1.20}$$

so that

$$\alpha_{HA} + \alpha_{A-} = 1 \tag{1.21}$$

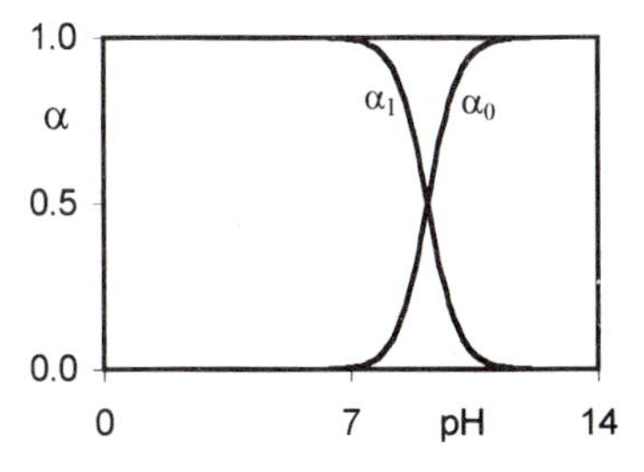

Fig. 1.2 The concentration fractions α_1 and α_0 (as a function of pH) of the acidic and basic forms, NH_4^+ and NH_3 respectively, of ammonia, $pK_a = 9.24$.

We can avoid superscripted subscripts, and at the same time make the formulas more generally applicable, by using a simpler notation, in which we denote the species by the number of attached, dissociable protons. Thus we will replace α_{HA} by α_1, and α_{A-} by α_0.

Upon combining eqns (1.4) and (1.18) we obtain

$$\alpha_1 = \frac{[HA]}{C} = \frac{[HA]}{[HA]+[A^-]} = \frac{[H^+][A^-]/K_a}{[H^+][A^-]/K_a + [A^-]}$$

$$= \frac{[H^+]/K_a}{[H^+]/K_a + 1} = \frac{[H^+]}{[H^+]+K_a} \tag{1.22}$$

Likewise we obtain

$$\alpha_0 = \frac{[A^-]}{C} = \frac{[A^-]}{[HA]+[A^-]} = \frac{[A^-]}{[H^+][A^-]/K_a + [A^-]}$$

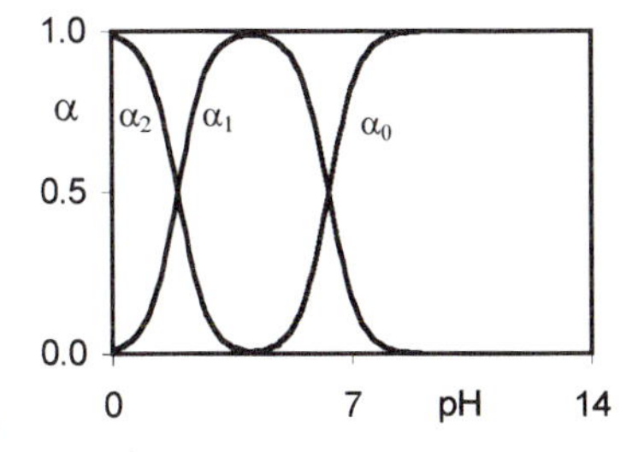

Fig. 1.3 The concentration fractions α_2, α_1 and α_0 of maleic acid, H_2Ma ($pK_{a1} = 1.91$, $pK_{a2} = 6.33$) and its partially and fully deprotonated forms, HMa^- and Ma^{2-}.

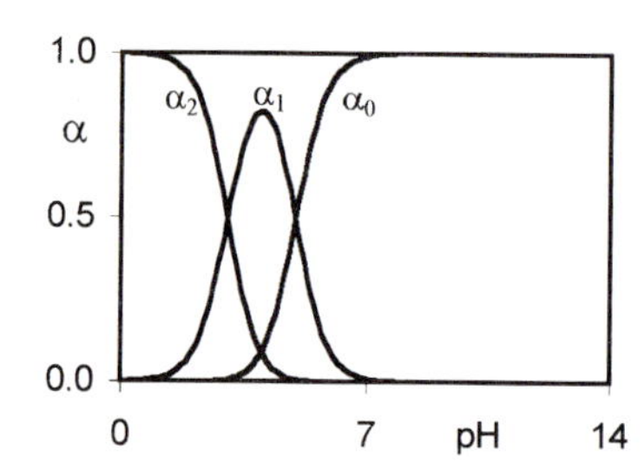

Fig. 1.4 The concentration fractions α_2, α_1 and α_0 of fumaric acid, H_2Fu (pK_{a1} = 3.05, pK_{a2} = 4.49) and its partially and fully deprotonated forms, HFu^- and Fu^{2-}.

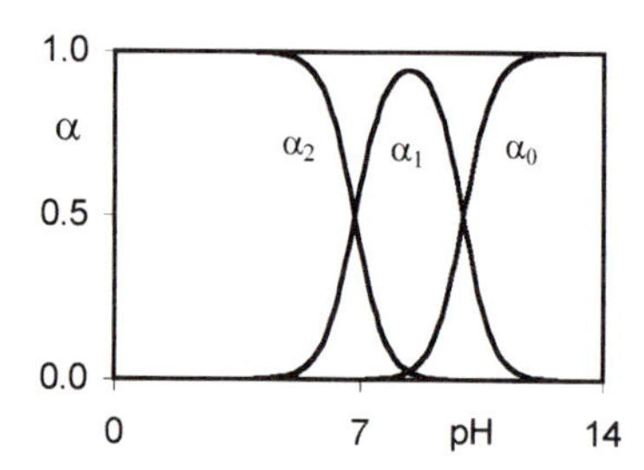

Fig. 1.5 The concentration fractions α_2, α_1 and α_0 of ethylenediamine, Ed (pK_{a1} = 6.85, pK_{a2} = 9.93) and its partially and fully protonated forms, HEd^+ and H_2Ed^{2+}.

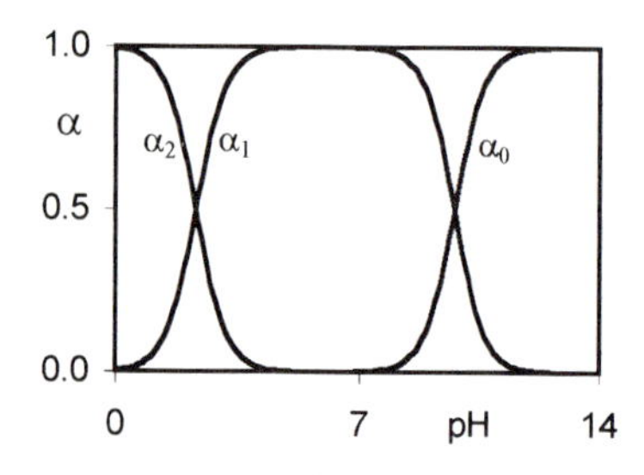

Fig. 1.6 The concentration fractions α_2, α_1 and α_0 of glycine, HGly (pK_{a1} = 2.35, pK_{a2} = 9.78) and its protonated and deprotonated forms, H_2Gly^+ and Gly^- respectively.

$$= \frac{1}{[H^+]/K_a + 1} = \frac{K_a}{[H^+] + K_a} \tag{1.23}$$

We will use concentration fractions α throughout this primer. It is useful to keep in mind that they are independent of the total analytical concentration C, but are explicit functions of $[H^+]$ as well as of the acid dissociation constant(s) K_a.

For diprotic acids we have, similarly,

$$\alpha_2 = \frac{[H_2A]}{C} = \frac{[H^+]^2}{[H^+]^2 + [H^+]K_{a1} + K_{a1}K_{a2}} \tag{1.24}$$

$$\alpha_1 = \frac{[HA^-]}{C} = \frac{[H^+]K_{a1}}{[H^+]^2 + [H^+]K_{a1} + K_{a1}K_{a2}} \tag{1.25}$$

$$\alpha_0 = \frac{[A^{2-}]}{C} = \frac{K_{a1}K_{a2}}{[H^+]^2 + [H^+]K_{a1} + K_{a1}K_{a2}} \tag{1.26}$$

where $K_{a1} \geq K_{a2}$. Likewise we obtain for triprotic acids

$$\alpha_3 = \frac{[H_3A]}{C} = \frac{[H^+]^3}{[H^+]^3 + [H^+]^2 K_{a1} + [H^+]K_{a1}K_{a2} + K_{a1}K_{a2}K_{a3}} \tag{1.27}$$

$$\alpha_2 = \frac{[H_2A^-]}{C} = \frac{[H^+]^2 K_{a1}}{[H^+]^3 + [H^+]^2 K_{a1} + [H^+]K_{a1}K_{a2} + K_{a1}K_{a2}K_{a3}} \tag{1.28}$$

$$\alpha_1 = \frac{[HA^{2-}]}{C} = \frac{[H^+]K_{a1}K_{a2}}{[H^+]^3 + [H^+]^2 K_{a1} + [H^+]K_{a1}K_{a2} + K_{a1}K_{a2}K_{a3}} \tag{1.29}$$

$$\alpha_0 = \frac{[A^{3-}]}{C} = \frac{K_{a1}K_{a2}K_{a3}}{[H^+]^3 + [H^+]^2 K_{a1} + [H^+]K_{a1}K_{a2} + K_{a1}K_{a2}K_{a3}} \tag{1.30}$$

where $K_{a1} \geq K_{a2} \geq K_{a3}$.

Figures 1.1 through 1.6 illustrate the dependence of the concentration fractions α on pH.

1.4 Logarithmic concentration diagrams

The mass action law is expressed in terms of the ratio of products of concentrations. In order to display such relations it is convenient to convert them into logarithmic form, because the logarithm of a product is simply the sum of the individual logarithms, while the logarithm of a ratio is merely the difference of the corresponding logarithms. Upon taking logarithms, the numerator and denominator in eqn (1.3) can be rewritten as a log[A] + b log[B] + c log[C] +… and p log[P] + q log[Q] + r log[R] + …

Just as pH is often a more useful quantity than $[H^+]$, double-logarithmic graphical representations can provide a useful overview of acid–base equilibria, because the logarithmic representation allows us to follow the concentrations of all species over a wide range of values, with uniform relative precision. In this primer we will often use plots in which the concentrations of individual species are shown over ten orders of magnitude, and that of $[H^+]$ over fourteen orders. An additional advantage of such *logarithmic concentration diagrams* is that they contain mostly linear segments of integer slopes. In this primer they will be used extensively, as an aid both in visualizing complicated equilibria and in making pH calculations.

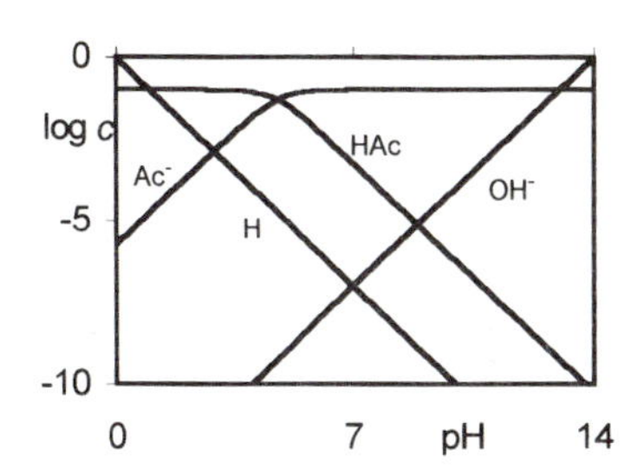

Fig. 1.7a The logarithmic concentration diagram shows the logarithms of the concentrations of the species H^+, HAc, Ac^-, and OH^- as a function of pH) for acetic acid (pK_a = 4.76) at a total analytical concentration C = 0.1 M.

In such a diagram we plot the logarithm of the concentrations of all relevant species involved in the equilibrium as a function of pH, which is itself the (negative) logarithm of the proton concentration. Figure 1.7a shows the logarithmic concentration diagram for acetic acid and/or acetate anions, with a pK_a of 4.76, at a total analytical concentration of 0.1 M. The vertical scale is labeled as log c, where c denotes the concentration of the particular species depicted.

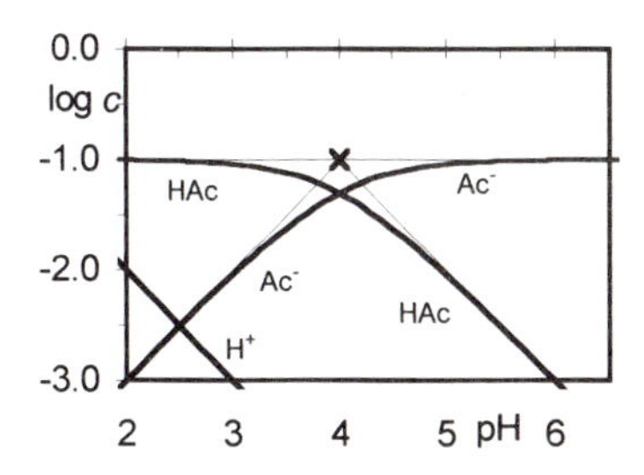

Fig. 1.7b Detail of Fig. 1.7a for acetic acid (pK_a = 4.76) at C = 0.1 M, showing the extension of the linear asymptotes (thin lines) to the system point (at the coordinates pH = pK_a, log c = log C).

Figure 1.7b contains four lines. The simplest of these is the line representing H^+ that goes through the top left–hand corner of the graph. It is a straight line of slope –1 because we plot log $[H^+]$ versus pH = – log $[H^+]$. It is most readily drawn by recognizing that that the line for H^+ passes through the points pH = 0, log c = 0 and pH = 10, log c = –10.

The line for OH^- is also a straight one, and passes through the top right-hand corner. In this primer we will assume, for the sake of convenience, the round value of 10^{-14} for K_w, see eqn. (1.6), which is its value at 25°C. The line for OH^- then passes through the points pH = 14, pc = 0, and pH = 4, log c = –10, and crosses the straight line for H^+ at pH = 7, log c = –7.

The line representing HA is not straight, although it has two straight–line asymptotes. At pH « pK_a, eqns (1.19) and (1.22) combine to

$$[\mathrm{HA}] = C\,\alpha_1 = \frac{C\,[\mathrm{H}^+]}{[\mathrm{H}^+]+K_a} \approx C \qquad \text{for } [\mathrm{H}^+] \gg K_a \tag{1.31}$$

so that pc = pC. At pH » pK_a, we instead obtain

$$[\mathrm{HA}] = C\,\alpha_1 = \frac{C\,[\mathrm{H}^+]}{[\mathrm{H}^+]+K_a} \approx \frac{C\,[\mathrm{H}^+]}{K_a} \qquad \text{for } [\mathrm{H}^+] \ll K_a \tag{1.32}$$

which in the logarithmic concentration diagram yields a line of slope –1, because log c = log $[H^+]$ + log C – log K_a = – pH + log C – log K_a. Finally, at pH = pK_a we have

$$[\mathrm{HA}] = C\,\alpha_1 = \frac{C\,[\mathrm{H}^+]}{[\mathrm{H}^+]+K_a} = C/2 \qquad \text{for } [\mathrm{H}^+] = K_a \tag{1.33}$$

so that log c = log C – log 2 ≈ log C – 0.30 because log 2 = 0.30103…

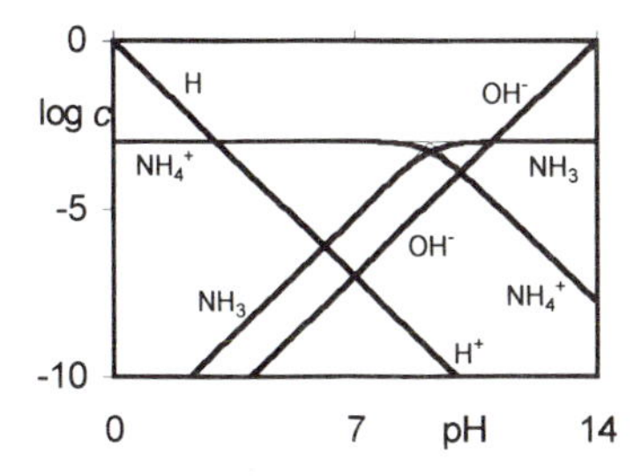

Fig. 1.8 The logarithmic concentration diagram for the monoprotic acid-base pair ammonium / ammonia (pK_a = 9.24) at a total analytical concentration C of 1 mM.

For A^- we find, analogously, a line with two linear asymptotes. At a pH much smaller than pK_a we have

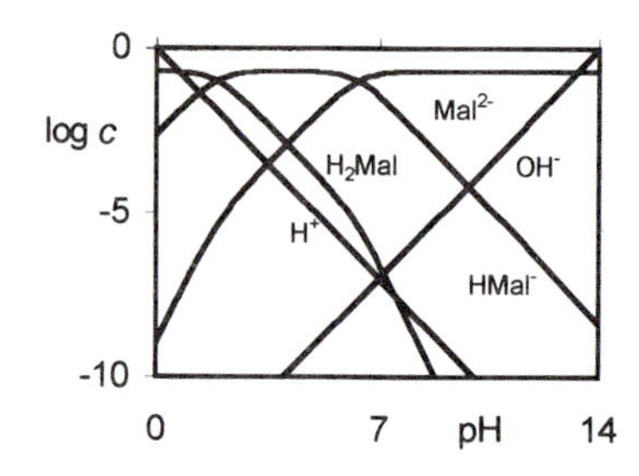

Fig. 1.9 The logarithmic concentration diagram for maleic acid (pK_{a1} = 1.91, pK_{a2} = 6.33) at a total analytical concentration C of 0.2 M. In this case the pK_as are sufficiently far apart, so that the behavior near one pK_a is only weakly affected by the presence of the other pK_a.

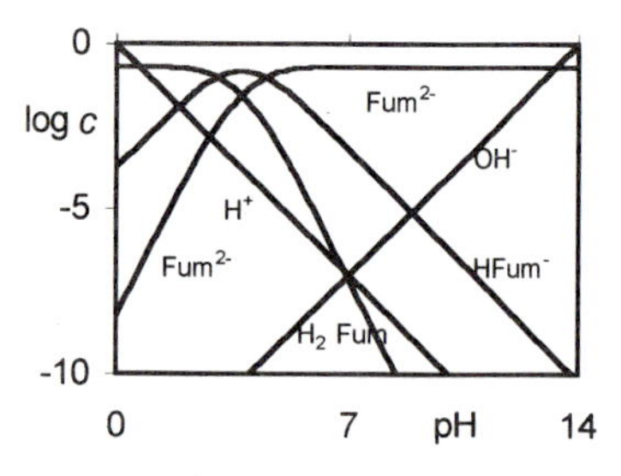

Fig. 1.10 The logarithmic concentration diagram for fumaric acid (pK_{a1} = 3.05, pK_{a2} = 4.49) at a total analytical concentration C of 0.2 M. In this case the two pK_as are close together.

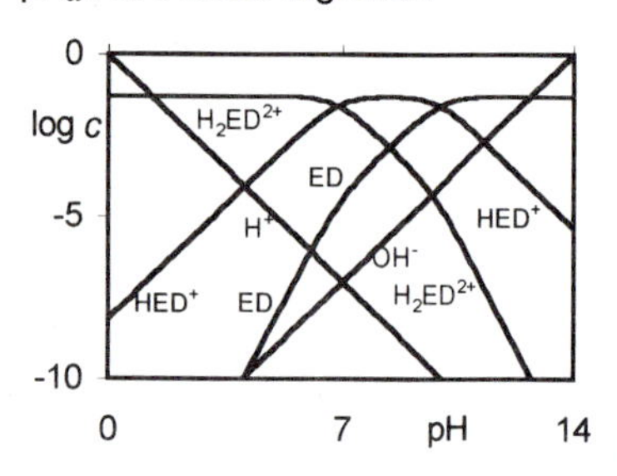

Fig. 1.11 The logarithmic concentration diagram for ethylenediamine (pK_{a1} = 6.85, pK_{a2} = 9.93) at a total analytical concentration C of 0.05 M.

$$[\mathrm{A}^-] = C\alpha_0 = \frac{C\,K_a}{[\mathrm{H}^+]+K_a} \approx \frac{C\,K_a}{[\mathrm{H}^+]} \qquad \text{for } [\mathrm{H}^+] \gg K_a \tag{1.34}$$

while, at pH values much larger than pK_a, we find

$$[\mathrm{A}^-] = C\alpha_0 = \frac{C\,[\mathrm{H}^+]}{[\mathrm{H}^+]+K_a} \approx C \qquad \text{for } [\mathrm{H}^+] \ll K_a \tag{1.35}$$

and, at pH = pK_a,

$$[\mathrm{A}^-] = C\alpha_0 = \frac{C\,[\mathrm{H}^+]}{[\mathrm{H}^+]+K_a} = C/2 \qquad \text{for } [\mathrm{H}^+] = K_a \tag{1.36}$$

As can be seen in Fig. 1.7b, the linear asymptotes extrapolate to a single point, with the coordinates pH = pK_a, pc = pC = – log C. That intersection of the extrapolated linear sections is called the *system point*, and plays a central role in the graphical representation.

Figure 1.8 shows a similar diagram for ammonia (pK_a = 9.24) at a total analytical concentration of 1 mM. Upon comparison of Figs. 1.7 and 1.8 with Figs. 1.1 and 1.2 respectively we note that the logarithmic concentration diagram has simpler curve shapes (with linear asymptotes of integer slopes), encompasses a much larger concentration range (because of its logarithmic compression), and shows the actual concentrations of all species involved in the acid-base equilibria, including those of H^+ and OH^-.

For a diprotic acid and its acid and normal salt, again at a total analytical concentration C, the logarithmic concentration diagram contains five species: H^+, OH^-, H_2A, HA^-, and A^{2-}. The representation of H^+ and OH^- in Figs. 1.9 through 1.12 is the same as in Figs. 1.7a and 1.8, but for the species H_2A, HA^-, and A^{2-} we now have

$$[\mathrm{H_2A}] = C\alpha_2 = \frac{C[\mathrm{H}^+]^2}{[\mathrm{H}^+]^2 + [\mathrm{H}^+]K_{a1} + K_{a1}K_{a2}} \tag{1.37}$$

$$[\mathrm{HA}^-] = C\alpha_1 = \frac{C[\mathrm{H}^+]K_{a1}}{[\mathrm{H}^+]^2 + [\mathrm{H}^+]K_{a1} + K_{a1}K_{a2}} \tag{1.38}$$

$$[\mathrm{A}^{2-}] = C\alpha_0 = \frac{C\,K_{a1}K_{a2}}{[\mathrm{H}^+]^2 + [\mathrm{H}^+]K_{a1} + K_{a1}K_{a2}} \tag{1.39}$$

which again yield lines with linear asymptotes. Moreover, when the pK_a-values are sufficiently far apart, these lines have a near–linear region between pH = pK_{a1} and pH = pK_{a2}. For example, at pH much smaller than pK_{a1}, so that $[\mathrm{H}^+] \gg K_{a1}$, eqn (1.37) reduces to $[\mathrm{H_2A}] \approx C$, which corresponds to a horizontal line in the diagram. On the other hand, for $[\mathrm{H}^+] \gg K_{a2}$ we find $[\mathrm{H_2A}] \approx C\,[\mathrm{H}^+]^2 / K_{a1}\,K_{a2}$. This shows in the diagram as a line of slope –2, since $\log\,[\mathrm{H_2A}] \approx -2\,\mathrm{pH} + \log C - \mathrm{p}K_{a1} - \mathrm{p}K_{a2}$, so that $\mathrm{d}\log c\,/\,\mathrm{d\,pH} \approx -2$.

For triprotic acids we have, likewise,

$$[H_3A] = C\alpha_3 = \frac{C[H^+]^3}{[H^+]^3 + [H^+]^2 K_{a1} + [H^+]K_{a1}K_{a2} + K_{a1}K_{a2}K_{a3}} \tag{1.40}$$

$$[H_2A^-] = C\alpha_2 = \frac{C[H^+]^2 K_{a1}}{[H^+]^3 + [H^+]^2 K_{a1} + [H^+]K_{a1}K_{a2} + K_{a1}K_{a2}K_{a3}} \tag{1.41}$$

$$[HA^{2-}] = C\alpha_1 = \frac{C[H^+]K_{a1}K_{a2}}{[H^+]^3 + [H^+]^2 K_{a1} + [H^+]K_{a1}K_{a2} + K_{a1}K_{a2}K_{a3}} \tag{1.42}$$

$$[A^{3-}] = C\alpha_0 = \frac{C\,K_{a1}K_{a2}K_{a3}}{[H^+]^3 + [H^+]^2 K_{a1} + [H^+]K_{a1}K_{a2} + K_{a1}K_{a2}K_{a3}} \tag{1.43}$$

as illustrated in double-logarithmic form in Figs. 1.13 and 1.14. Again, the lines have linear asymptotes at pH « pK_{a1} and pH » pK_{a3} . Moreover, when the pK_as are sufficiently far apart, as in Fig. 1.13, the diagram also shows linear sections in the intermediate pH ranges, between those pK_as.

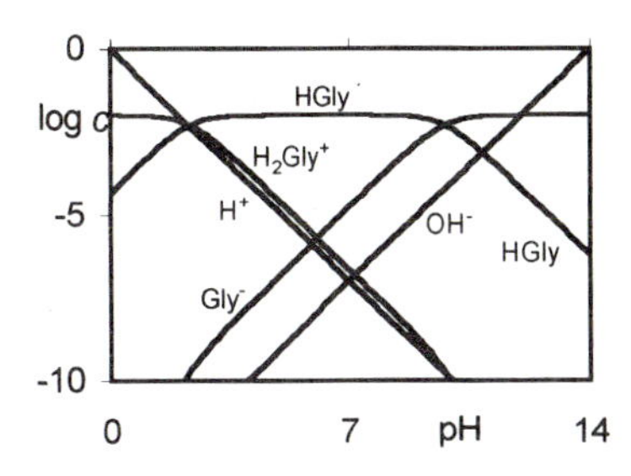

Fig. 1.12 The logarithmic concentration diagram for glycine (pK_{a1} = 2.35, pK_{a2} = 9.78) at a total analytical concentration C of 0.01 M.

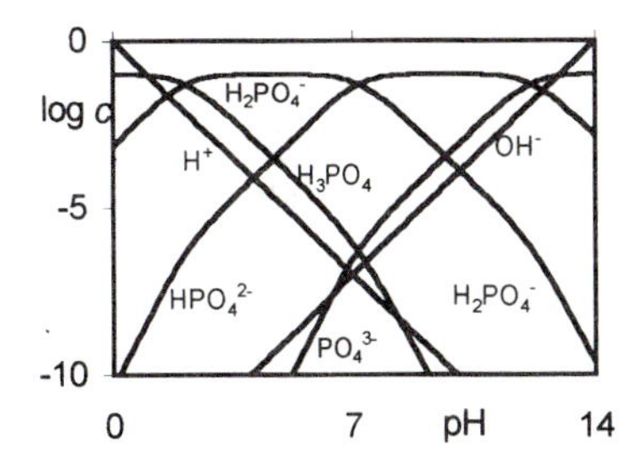

Fig. 1.13 The logarithmic concentration diagram for phosphoric acid (pK_{a1} = 2.15, pK_{a2} = 7.20, pK_{a3} = 12.15) at a total analytical concentration C of 0.1 M.

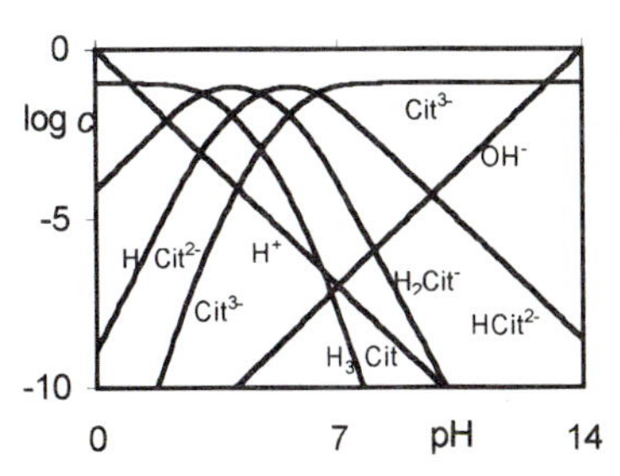

Fig. 1.14 The logarithmic concentration diagram for citric acid (pK_{a1} = 3.13, pK_{a2} = 4.76, pK_{a3} = 6.40) at a total analytical concentration C of 0.1 M.

1.5 The proton condition

In order to calculate the concentrations of all species involved in the acid–base equilibrium, we must know the equilibrium constant(s) K_a as well as the analytical concentration C of the acid. We also need to consider two *conservation laws*, namely those for mass and charge. These laws state that neither mass nor charge can just appear or disappear in a chemical reaction. These two laws hold quite generally for all kinds of chemical reactions; in this section we will illustrate their application to acid–base equilibria.

Consider the aqueous solution of C moles per liter of a weak acid HA. Water may partially dissociate according to $H_2O \rightleftharpoons H^+ + OH^-$, while a fraction of the acid may similarly dissociate through $HA \rightleftharpoons H^+ + A^-$. The resulting solution species are then, H_2O, HA, H^+, A^-, and OH^-, and the conservation laws describe the constraints on these concentrations. In this case we know that the total analytical concentration of the acid is C, so that we have the *mass balance equation* which, in this case, is merely the original definition of C,

$$[HA] + [A^-] = C \tag{1.44}$$

Likewise, electroneutrality of the solution as a whole requires that the sum of all the positive and negative charges must add up to zero, or

$$[H^+] = [A^-] + [OH^-] \tag{1.45}$$

which is the appropriate *charge balance equation*. In general, the conservation of mass will give rise to at least one mass balance equation for each total analytical concentration. In mixtures, we may therefore have a multitude of mass balance equations. In contrast, charge conservation always leads to *only one* charge balance equation, which must contain the concentrations of all ionic species in a given solution.

As our second example we consider an aqueous solution of C moles of the salt NaA of the weak acid HA and the strong base NaOH. We can now write two mass balance equations, namely eqn (1.44) and

$$[Na^+] = C \tag{1.46}$$

plus the charge balance equation

$$[H^+] + [Na^+] = [A^-] + [OH^-] \tag{1.47}$$

The charge balance equation now contains the concentration $[Na^+]$, which is somewhat peripheral to acid–base problems. It is often convenient to eliminate such non–essential terms from the charge balance equation. In the above example, we combine the two mass balance equations as $[Na^+] = [HA] + [A^-]$, then substitute this into eqn (1.47) to obtain

$$[H^+] + [HA] = [OH^-] \tag{1.48}$$

This result can be rationalized in a quite different way. Consider the initial ingredients that go into making the aqueous solution of C M NaA as the neutral species H_2O and the ionic salt NaA. We now count solution species that, with respect to those (somewhat hypothetical) initial components, have acquired or lost protons. We will denote water molecules that have gained a proton as H^+, and salt anions that have gained a proton as HA. (Instead of H^+ we might be tempted to write H_3O^+, but in this primer we will delete integer units of H_2O from our chemical equations in order to simplify them. At any rate, we often do not know the precise chemical composition of hydrated species, and the latter, usually being a statistical average, may not even be expressible in terms of an integer number of hydration molecules.)

On the other side of the ledger, water can have lost a proton, in which case it has become OH^-, while neither Na^+ nor A^- has any protons to lose. Consequently, eqn (1.48) can be considered to be a *proton balance equation* or, for short, a *proton condition*, expressing the balance between proton gainers and losers. Usually, the proton condition is the only combination of the mass and charge balance equations needed to solve pH problems. Moreover, the proton condition can often be written down simply by accounting for all species that gain and lose protons. In this primer, *the proton condition will be the primary quantity for computing the pH of a solution.*

If there is any ambiguity in formulating the proton condition, the rule is that the proton condition should *not* contain the starting species, where we consider acids and bases as undissociated, salts as dissociated. For example, the proton condition for HA should not contain [HA], as in eqn (1.45), while the proton condition for NaA should contain neither $[Na^+]$ nor $[A^-]$, see eqn (1.48). This rule is automatically assured when we consider proton gainers and losers.

Because of the central position of the proton condition in this book, we will here give some additional examples of its formulation. For an aqueous solution of NH_4Cl we have the mass balance equations $[NH_4^+] + [NH_3] = C$ and $[Cl^-] = C$, the charge balance equation $[H^+] + [NH_4^+] = [Cl^-] + [OH^-]$, and the proton condition $[H^+] = [NH_3] + [OH^-]$.

For ammonia, NH_3, we would instead have the mass balance $[NH_4^+]$ + $[NH_3]$ = C, and the charge balance $[H^+]$ + $[NH_4^+]$ = + $[OH^-]$, which now doubles as the proton condition.

Applying these same bookkeeping considerations to the diprotic acid H_2A yield the mass and charge balance equations

$$[H_2A] + [HA^-] + [A^{2-}] = C \tag{1.49}$$

$$[H^+] = [HA^-] + 2[A^{2-}] + [OH^-] \tag{1.50}$$

where the factor 2 in the charge balance equation (1.50) reflects the divalent nature of A^{2-}. Note that eqn (1.50) again doubles as the proton condition.

For the acid salt NaHA, as in sodium hydrogen phthalate, we have two mass balances, eqn (1.46) and (1.49), and one charge balance relation,

$$[H^+] + [Na^+] = [HA^-] + 2[A^{2-}] + [OH^-] \tag{1.51}$$

while the proton condition is

$$[H^+] + [H_2A] = [A^{2-}] + [OH^-] \tag{1.52}$$

as can be derived from eqns (1.46), (1.49), and (1.51), or (simpler) found by considering the proton gainers and losers among the starting species H_2O, Na^+, and HA^-.

For the disodium salt Na_2A we have the mass balance equations (1.49) and $[Na^+] = 2C$, the charge balance relation (1.51), and the proton condition

$$[H^+] + 2[H_2A] + [HA^-] = [OH^-] \tag{1.53}$$

reflecting the fact that the initial species are H_2O, Na^+, and A^{2-}, so that A^{2-} can gain either one proton (forming HA^-) or two (to yield H_2A).

As our last example we consider a C M solution of $Na(NH_4)_2PO_4$. We have the mass balance relations $[H_3PO_4] + [H_2PO_4^-] + [HPO_4^{2-}] + [PO_4^{3-}] = C$, $[Na^+] = C$, and $[NH_3] + [NH_4^+] = 2C$. Combining these with the charge balance equation $[H^+] + [Na^+] + [NH_4^+] = [H_2PO_4^-] + 2[HPO_4^{2-}] + 3[PO_4^{3-}] + [OH^-]$ and, eliminating the terms $[Na^+]$, $[NH_4^+]$, and $[PO_4^{3-}]$ that do not figure among the proton gainers or losers, yields the proton condition $[H^+] + 3[H_3PO_4] + 2[H_2PO_4^-] + [HPO_4^{2-}] = [NH_3] + [OH^-]$. In Table 1.1 we list the proton conditions for ortho-phosphoric acid and its sodium and ammonium salts. The reader is invited to verify these, either by inspection, or by combining the relevant mass and charge balance relations.

In connection with the laws of conservation of mass and charge, and the associated proton balance, it is useful to point out that the conservation laws are strictly *bookkeeping* devices, counting particles, charges, and protons respectively, and are therefore not subject to activity corrections. The second is that these laws only apply to so-called 'closed' systems, i.e., to regions of space that do not exchange material with their surroundings. In Chapter 3 we will see how they can be modified to apply during a titration, i.e., when a titrant is added to the sample.

Table 1.1 The proton conditions for *o*-phosphoric acid and some of its salts

H_3PO_4:	$[H^+] = [H_2PO_4^-] + 2[HPO_4^{2-}] + 3[PO_4^{3-}] + [OH^-]$
NaH_2PO_4:	$[H^+] + [H_3PO_4] = [HPO_4^{2-}] + 2[PO_4^{3-}] + [OH^-]$
$NH_4H_2PO_4$:	$[H^+] + [H_3PO_4] = [HPO_4^{2-}] + 2[PO_4^{3-}] + [NH_3] + [OH^-]$
Na_2HPO_4:	$[H^+] + 2[H_3PO_4] + [H_2PO_4^-] = [PO_4^{3-}] + [OH^-]$
$NaNH_4HPO_4$:	$[H^+] + 2[H_3PO_4] + [H_2PO_4^-] = [PO_4^{3-}] + [NH_3] + [OH^-]$
$(NH_4)_2HPO_4$:	$[H^+] + 2[H_3PO_4] + [H_2PO_4^-] = [PO_4^{3-}] + [NH_3] + [OH^-]$
Na_3PO_4:	$[H^+] + 3[H_3PO_4] + 2[H_2PO_4^-] + [HPO_4^{2-}] = [OH^-]$
$Na_2NH_4PO_4$:	$[H^+] + 3[H_3PO_4] + 2[H_2PO_4^-] + [HPO_4^{2-}] = [NH_3] + [OH^-]$
$Na(NH_4)_2PO_4$:	$[H^+] + 3[H_3PO_4] + 2[H_2PO_4^-] + [HPO_4^{2-}] = [NH_3] + [OH^-]$
$(NH_4)_3PO_4$:	$[H^+] + 3[H_3PO_4] + 2[H_2PO_4^-] + [HPO_4^{2-}] = [NH_3] + [OH^-]$

For mixtures, finding the proton condition may not always be so intuitive. In such a case one always can (and usually should) go back to the mass and charge balance relations in order to derive the proton condition. Moreover, in that case we may occasionally find several, alternative forms of the proton condition. While these will be *mathematically* equivalent, typically one of them will be the more useful for our purposes. This most useful form will have the steepest pH dependencies for the proton gainers and proton losers respectively. This will be illustrated in Chapter 2.

2 Numerical solutions

In general, we can compute the pH of any aqueous solution by solving its proton condition. In the previous Chapter we have already shown how to write the individual concentration terms occurring in the proton condition as explicit functions of $[H^+]$; after that, finding the unique solution to the proton condition merely requires well-known mathematical methods. Consequently, in a purely formal sense, the problem is mathematically well-posed and straightforward, and would require little further discussion. Why, then, devote an entire Chapter to it?

While it is possible to view the problem as a purely mathematical one, few chemists feel comfortable with such an approach. Moreover, the purely mathematical solution is not without difficulties, a consequence of the large range of $[H^+]$, which can vary over some 15 orders of magnitude.

In the present Chapter we will instead explore how to find the solution by combining graphical and mathematical methods, with as much emphasis as possible on the graphs, because the latter display the relevant *chemical* information. In this context it is useful to remember that we require accurate but not really highly precise answers, because the results cannot be better than the input information used, which is usually available in the form of pK-values known only to about ±0.01. Moreover, the inherent uncertainties of ionic activity coefficients often result in uncertainties in the pH of the order of ±0.1 or more. Given these physical and chemical realities, the graphical method is usually all we need, because we can typically find the pH to within ±0.1 directly from the logarithmic concentration diagram, while simple refinements often provide pH values good to ±0.01. Only in the most complicated cases will we have no choice but to resort to purely mathematical methods.

This Chapter, then, will focus on the logarithmic concentration diagram, because it shows the relative magnitudes of the various terms that make up the proton condition, and therefore often allows us to find a satisfactory *approximate* solution without the help of a computer. Using the latter approach is often faster, and also provides a valuable overview of the composition of the solution.

We will treat the numerical solutions of pH problems in order of their increasing complexity, and of the sophistication and numerical prowess of the tools needed for their solution. First we will use the logarithmic concentration diagram to *visualize* the proton condition, because this will immediately show us what methods we need to use. In Section 2.2 we will then consider methods that need no tools beyond a piece of scrap paper and a pencil, and in section 2.3 those methods that require a simple calculator capable of computing square roots and taking logarithms. Only in Section 2.4 will we describe general methods based on relatively simple iterative methods, such as a Newton-Raphson algorithm. Because iterations are repetitive, such calculations are most readily performed on a computer, using generally available software such as a spreadsheet, or even on some of the available scientific pocket calculators. Finally we will briefly describe some even more general approaches used in commercial programs that solve an arbitrary number of equilibrium expressions and the associated mass and charge balance equations.

2.1 Visualizing the proton condition

In view of the central role played by the proton condition, we start by visualizing the proton condition. The proton condition of an aqueous solution always contains at least the term $[H^+]$ on its left-hand side (for proton gainers), and the term $[OH^-]$ on its right-hand side (for proton losers), since these terms reflect the autodissociation of the solvent. Other terms must be added as needed: for acids, there will be terms on the right-hand side to represent its partially or fully deprotonated forms, for bases there will be terms on the left-hand side for their partially or fully protonated forms, while acidic or basic salts contribute terms to both sides. In general, then, we have one or more terms on both sides of the equal sign, and we will visualize both in the logarithmic concentration diagram.

The proton condition is satisfied when the left- and right-hand sides are equal. This will occur at a particular, unique pH, and determining that pH will be the main focus of this Chapter.

When we separately display the left- and right-hand sides of the proton condition (i.e., the terms representing the proton gainers and proton losers respectively) we will get two lines. We will distinguish between *lines* and *straight lines*. Almost all lines discussed in this Chapter will either be straight lines or have major *sections* that, for all practical purposes, are straight, interconnected by regions with appreciable curvature.

At the pH where the two sides of the proton condition are equal, the lines representing proton gainers and losers must intersect. The problem of finding the pH of the solution is therefore equivalent to finding the *intersection* of the two lines associated with the (properly weighted) sum of the concentrations of the proton gainers and losers. The weighting involves integer positive coefficients (mostly ones, but sometimes twos, threes etc.) reflecting the number of protons lost or gained.

The logarithmic concentration diagram displays the concentrations of all solution species we need to consider, in logarithmic form. This form facilitates the representation of the mass action law, which contains the quotient of products of powers, because $\log \{[P]^p[Q]^q[R]^r.../[A]^a[B]^b[C]^c...\} = p \log[P] + q \log[Q] + r \log[R] + ... a \log[A] - b \log[B] - c \log[C] - ...$ On the other hand, summation of individual concentration terms, as occurs in the proton condition, is not so straightforward to represent in a logarithmic diagram, because $\log(x+y)$ cannot be expressed as a simple function of $\log(x)$ and $\log(y)$.

The logarithm of a sum, $\log(x+y+...)$, depends very strongly on the dominant term(s) in that sum, and we can often ignore minor terms. To illustrate this, consider the logarithm of the sum $1 + 0.1 + 0.01 + 0.001 + 0.0001$

+ 0.00001 = 1.11111, where successive terms differ by an order of magnitude. We have log(1.1111) = 0.045753. When we only need the answer to two decimal places, we see that we do not need to consider all six terms: log(1) = 0.000000, log(1.1) = 0.041393, log(1.11) = 0.045323, log(1.111) = 0.045753, and log(1.1111) = 0.045753. For a result good to within ±0.01 we only need to consider the first two terms, which yield 0.04; if we want to be extra careful, we use the first three terms, which produce a result to within ±0.001, or in the above example 0.04_6 which, after rounding, produces 0.05. In general, for answers good to ±0.01, individual terms in the sum that are less than 1% of the leading term(s) can be neglected, and even terms that contribute less than 10% of the total sum are of relatively little consequence. This practical consideration often leads to a great reduction in the complexity of the problem.

Note that finding the pH is unlike computing the value of π, an abstract mathematical quantity meaningful to millions of decimal places. The limitation is *not* one of present technology: many commercial pH meters already measure to ±0.001 pH units, and determining the pH to four decimal places, i.e., to ±6μV, is no great technological feat. However, the *interpretation* of such measurements is subject to a more fundamental constraint that we will address in Chapter 7 of this book.

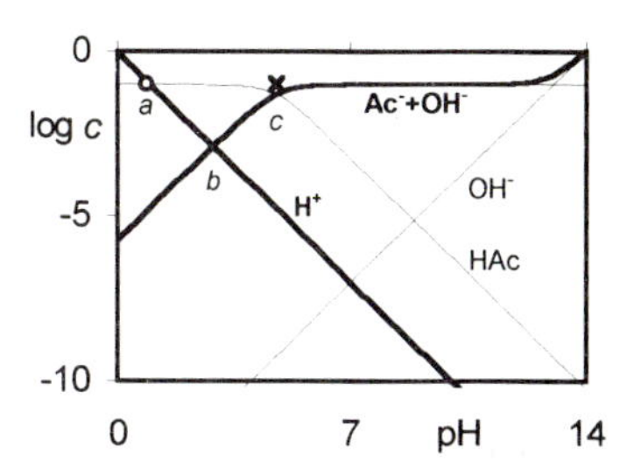

Fig. 2.1 The logarithmic concentration diagram of 0.1 M acetic acid (pK_a = 4.76), showing the two parts (proton gainers and losers) of its proton condition, the system point (cross), and an auxiliary point (open circle on the H^+ line at log c = log C) used to compute the pH.

After all these introductory comments we will now illustrate the visualization of the proton condition, using a few simple examples. Figure 2.1 shows the logarithmic concentration diagram for 0.1 M acetic acid, HAc. The pK_a of acetic acid is 4.76, and we therefore draw Fig. 2.1 with its system point at pH = 4.76, log c = –1. The two sides of the proton condition, $[H^+] = [A^-] + [OH^-]$, are shown as heavy lines for $\log[H^+]$ and $\log\{[A^-]+[OH^-]\}$ respectively. At high pH, where $[A^-] \ll [OH^-]$, the line for $\log\{[A^-]+[OH^-]\}$ follows that for $\log[OH^-]$. At low pH, the situation is reversed, because $[A^-] \gg [OH^-]$, so that $\log\{[A^-]+[OH^-]\} \approx \log[A^-]$. In between, there is a relatively small transition region where $[A^-]$ and $[OH^-]$ are of comparable magnitudes, so that $\log\{[A^-]+[OH^-]\}$ is substantially different from either $\log[A^-]$ or $\log[OH^-]$. The most convenient point in this region is where $[A^-] = [OH^-]$, because there $\log\{[A^-]+[OH^-]\} = \log\{2\times[A^-]\} = \log 2 + \log[A^-] = 0.30 + \log[A^-]$, since $\log(2) = 0.301030 \approx 0.30$.

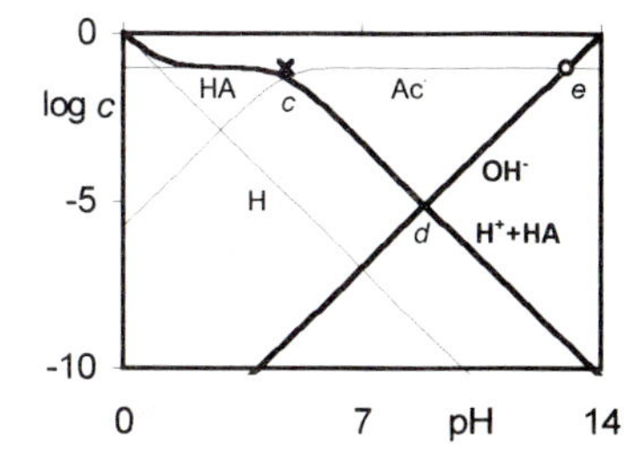

Fig. 2.2 The logarithmic concentration diagram of 0.1 M sodium acetate, showing the two parts of its proton condition, the system point (cross) and an auxiliary point (circle on the OH^- line at log c = log C).

Figure 2.2 shows the corresponding diagram for 0.1 M NaAc. This is essentially the same diagram, except that the proton condition for the salt is $[H^+] + [HA] = [OH^-]$, so that the heavy lines representing proton gainers and losers now show $\log\{[H^+]+[HA]\}$ and $\log[OH^-]$ respectively. Again, the line for $\log\{[H^+]+[HA]\}$ mostly follows either $\log[H^+]$ or $\log[HA]$, except in the small region where $[H^+] \approx [HA]$. At the point where $[H^+] = [HA]$, the line for $\log\{[H^+]+[HA]\}$ lies 0.30 above the values for $\log[H^+]$ and $\log[HA]$.

Finally, Figures 2.3 and 2.4 illustrate the corresponding diagrams for 0.5 μM HAc and 0.5 μM NaAc respectively. Note that the proton condition $[H^+] = [A^-] + [OH^-]$ for HAc is written the same way for 0.1 M and 10 μM acetic acid, but that the actual concentrations [HA] and $[A^-]$ are quite different in

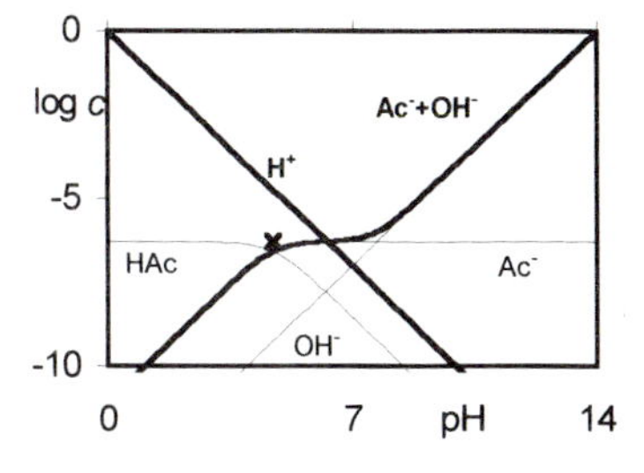

Fig. 2.3 The logarithmic concentration diagram of 0.5 μM acetic acid, with the two parts of its proton condition.

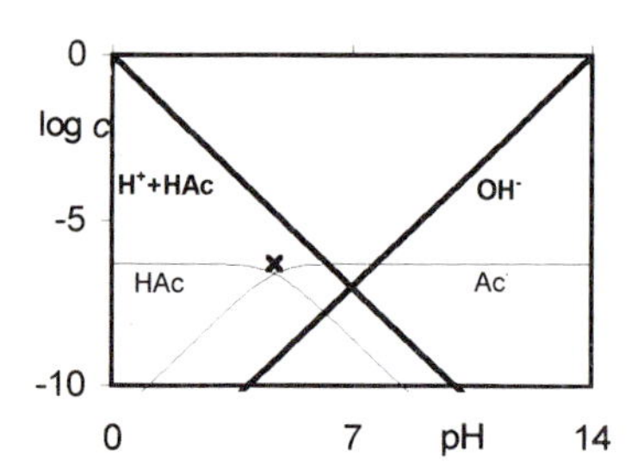

Fig. 2.4 The logarithmic concentration diagram of 0.5 μM sodium acetate, with the two parts of its proton condition.

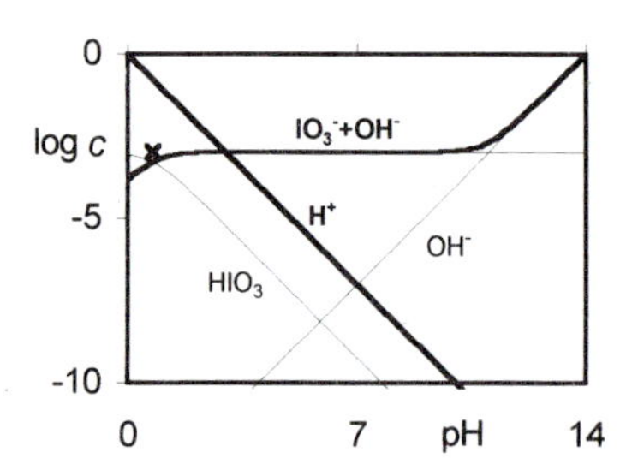

Fig. 2.5 The logarithmic concentration diagram of 1 mM iodic acid (HIO_3, pK_a = 0.77), with the two parts of its proton condition.

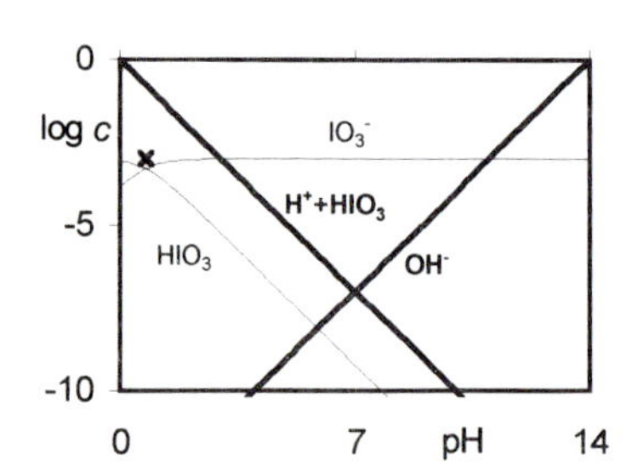

Fig. 2.6 The logarithmic concentration diagram of 1 mM sodium iodate, with the two parts of its proton condition.

these two cases, so that the lines for $\log[H^+]$ and $\log\{[A^-]+[OH^-]\}$ in Fig. 2.3 are quite different from those in Fig. 2.1. The same applies for a comparison of Figs. 2.2 and 2.4.

Note that all these examples illustrate that, in the double-logarithmic representation of a logarithmic concentration diagram, the sum of two intersecting straight lines is not a straight line itself, but a line that closely follows the larger of the two underlying straight lines. In purely mathematical terms we have $\log(a+b) \neq \log(a) + \log(b)$ but, for $a \gg b$, $\log(a+b) \approx \log(a)$, while $a \ll b$ leads to $\log(a+b) \approx \log(b)$.

2.2 Simple approximations

We now use these diagrams to compute the pH. In Figure 2.1 the proton condition for 0.1 M acetic acid, Eqn (1.45), is satisfied at the intersection of the lines for $[H^+]$ and $\{[A^-]+[OH^-]\}$ or their logarithms. In fact, at that intersection, $[A^-] > 10^{-3}$ M while $[OH^-] < 10^{-11}$ M, so that we can use the approximation $[Ac^-] + [OH^-] \approx [Ac^-]$ with an error less than 1 in 10^8, a pretty good approximation!

Now that we have found the appropriate approximation, $[H^+] = [Ac^-] + [OH^-] \approx [Ac^-]$, we use the diagram to compute the pH as follows. First we note that the lines for $\log[H^+]$ and $\log[Ac^-]$ are straight in the pH region of interest. Moreover, they have slopes of -1 and $+1$ respectively. In order to facilitate the calculation, we enter an *auxiliary point* (labeled a in Fig. 2.1) on the line for $[H^+]$ at the same height as the system point, i.e., at $c = C$. Therefore, the pH of the intersection, at the point labeled b, lies exactly halfway between points a and c, i.e., pH = (1 + 4.76) / 2 = 2.88. This corresponds to the simple formula $\text{pH} = (\text{p}C + \text{p}K_a) / 2$ (where the operator p denotes $-\log$), or to $[H^+] = \sqrt{CK_a}$, an equation we did *not* use.

Note that we have used the diagram as a *guide* for the calculation, but have nowhere actually read it, or relied on its precision. This calculation can therefore be made based on a crude sketch, on a drawing in sand on the beach, or on a purely mental image, and certainly requires neither a high level of draftsmanship, nor graph paper.

Similarly, Fig. 2.2 can be used to find the pH of sodium acetate (or any other salt of acetic acid with a strong base). In this case the proton condition is given by Eqn (1.48). At a pH above about 6, the line for [HAc] lies more than 3 logarithmic units above that for $[H^+]$, so that [HAc] is more than a thousand times larger than 10^3 $[H^+]$. Therefore we approximate Eqn (1.48) to $[HAc] \approx [OH^-]$. Consequently we find the intersection labeled d in Fig. 2.1, half-way between points c and e, where the auxiliary point e now lies on the line for $[OH^-]$ at $c = C$. Point c, the system point, has a pH of 4.76, and point e, at the intersection between the lines for Ac^- and OH^-, lies at pH 13, one logarithmic unit below and one logarithmic unit to the left of the top-right corner of the diagram. Consequently we find pH = (4.76 + 14 – $\log C$)/2 = (4.76 + 13) / 2 = 8.88. This result is equivalent to $\text{pH} = (\text{p}K_a + \text{p}K_w - \text{p}C) / 2$ or $[H^+] = \sqrt{K_aK_w/C}$. Again, we did not use that equation, but instead directly found the equivalent numerical answer.

Now we ask the same questions for 0.5 μM acetic acid, and for 0.5 μM NaAc. Figure 2.3 shows the logarithmic concentration diagram for the acid. Compared with Fig. 2.1, the system point for 0.5 μM HAc is shifted downwards by five logarithmic units, and the lines for [HAc] and $[Ac^-]$ are shifted downward accordingly. Again we look for the pH where the proton condition (1.45) is satisfied, i.e., for the intersection of $\log[H^+]$ with $\log\{[Ac^-]+[OH^-]\}$, and again we find that, at this intersection, $[Ac^-] + [OH^-] \approx [Ac^-]$ so that the proton condition $[H^+] = [Ac^-] + [OH^-]$ can be approximated by $[H^+] \approx [Ac^-]$. We now read the pH from the diagram as pH $\approx$ pC = 6.0. Note that we now have $[H^+] = C$ instead of $[H^+] = \sqrt{CK_a}$.

For 0.5 μM sodium acetate we likewise simplify the proton condition $[H^+] + [HAc] = [OH^-]$ to $[H^+] \approx [OH^-]$ so that pH $\approx$ 7.0. Again, this result does not obey the same mathematical relation as that for 0.1 M NaAc.

Figure 2.5 shows the logarithmic concentration diagram for 1 mM iodic acid, HIO_3, which has a pK_a of 0.77. Compared with Fig. 2.1, the system point now lies about four logarithmic units towards lower pH values (i.e., towards the left) and two logarithmic units towards lower concentrations (i.e., down), while the lines representing HIO_3 and IO_3^- follow suit. Again we look for the pH where the proton condition (1.45) is satisfied, i.e., for the intersection of $\log[H^+]$ with $\log\{[IO_3^-]+[OH^-]\}$, and again we find that, at this intersection, $[IO_3^-] + [OH^-] \approx [IO_3^-]$ so that the proton condition $[H^+] = [IO_3^-] + [OH^-]$ can be approximated by $[H^+] \approx [IO_3^-]$. We now read the pH from the diagram as pH = pC = 3.0.

For the corresponding salt, $NaIO_3$, we use the proton condition (1.48), i.e., $[H^+] + [HIO_3] = [OH^-]$. Since the diagram (Fig. 2.6) shows that $[HIO_3]$ is more than two orders of magnitude smaller than $[H^+]$ at all pH values (i.e., everywhere in the diagram), we simplify the proton condition to $[H^+] \approx [OH^-]$ and read off the diagram as pH = 7.0.

For 1 mM ammonia (NH_3 or, written in its hydrated form, NH_4OH), with a pK_a of 9.24, we now draw the logarithmic concentration diagram of Fig. 2.7. The proton condition reads $[H^+] + [NH_4^+] = [OH^-]$, and the diagram shows that we can simplify this, in the region around the intersection of $\log([H^+]+[NH_4^+])$ and $\log[OH^-]$, to $[NH_4^+] \approx [OH^-]$. We read off the graph that this point, labeled e, lies half-way between the system point and the point at pH = 14 – pC, – log(c) = – log(C), so that pH $\approx$ (9.24+11)/2 = 10.1.

For 1 mM NH_4Cl we find the proton condition $[H^+] = [NH_3] + [OH^-]$ which leads us to point b in Fig. 2.8. In that region, $[NH_3] \gg [OH^-]$, so that $[NH_3] + [OH^-] \approx [NH_3]$. Consequently we find the pH as halfway between system points and the point at pH = pC, – log(c) = – log(C), or pH $\approx$ (3.00+9.24)/2 = 6.1.

We now consider a few slightly more complicated cases. Fig. 2.9 shows the logarithmic concentration diagram for 0.1 M ammonium acetate, for which the proton condition reads $[H^+] + [HAc] = [NH_3] + [OH^-]$. In the region of the intersection of $\log([H^+]+[HAc])$ and $\log([NH_3]+[OH^-])$, near point c, we can simplify these expressions to $\log[HAc]$ and $\log[NH_3]$ respectively, and find pH = (4.76+9.24)/2 = 7.0. Note that the same diagram can be used to find the pH of HAc (at point a), NH_4Cl (point b), NaAc (point d), and NH_3 (point e), all at 0.1 M.

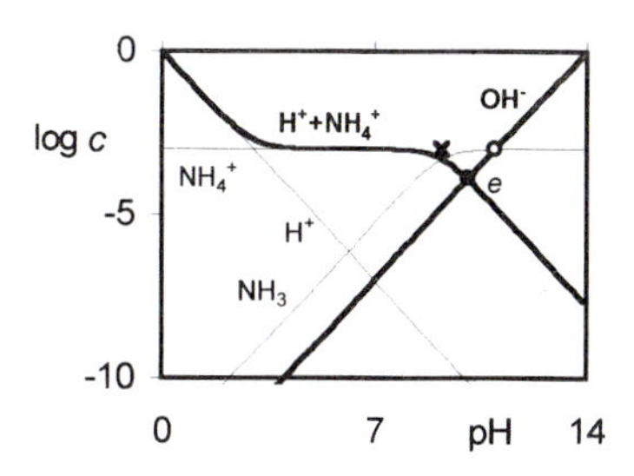

Fig. 2.7 The logarithmic concentration diagram of 1 mM ammonia, pK_a = 9.24.

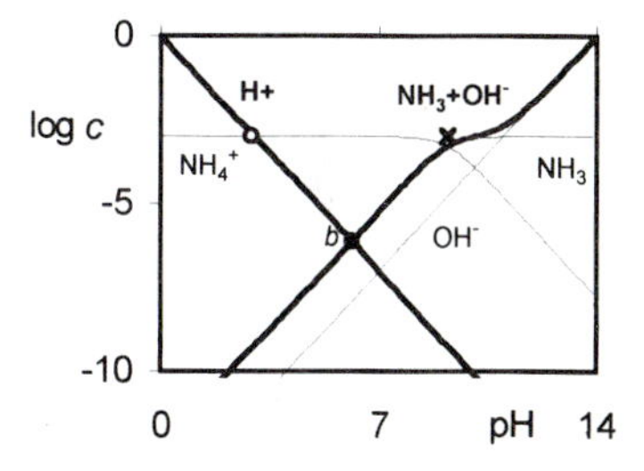

Fig. 2.8 The logarithmic concentration diagram of 1 mM ammonium chloride.

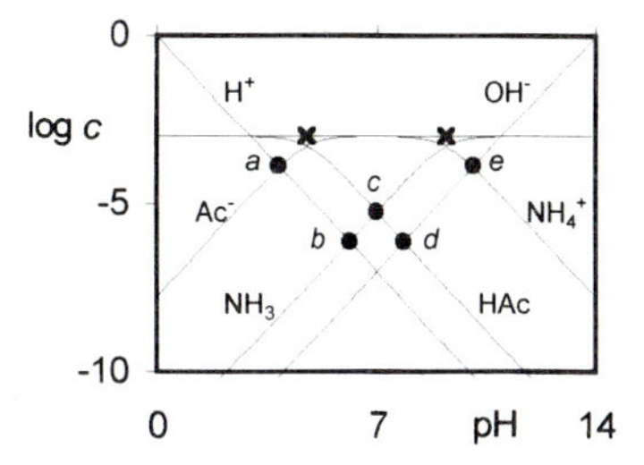

Fig. 2.9 The logarithmic concentration diagram of 1 mM ammonium acetate (point c), together with the pH-values for HAc (point a), $NaNH_4$ (point b), NaAc (point d), and NH_3 (point e), all at 1 mM.

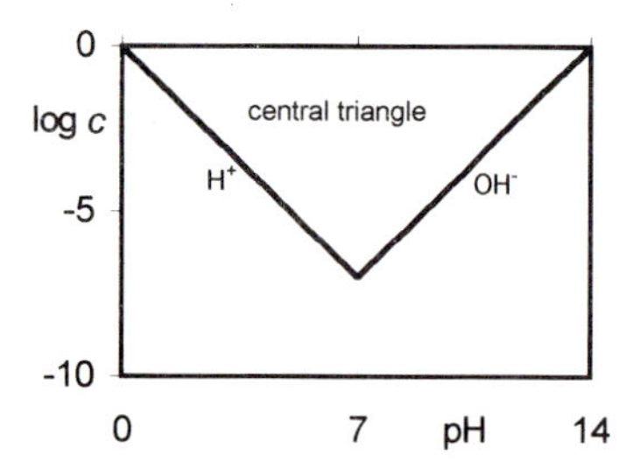

Fig. 2.10 The position of a system point with respect to the central triangle defines how an acid or base will behave.

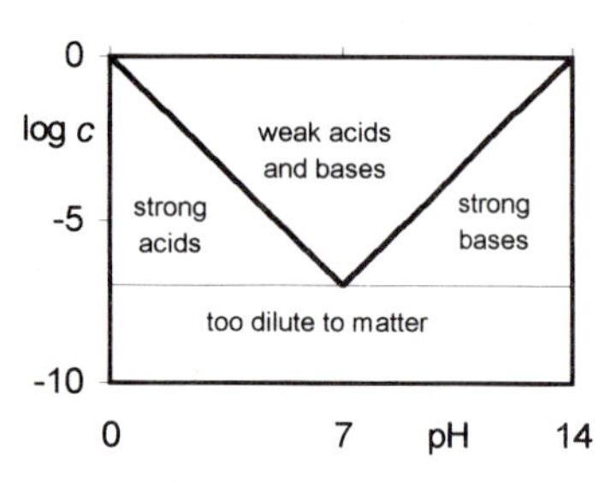

Fig. 2.11 Acids and bases act as weak ones when their system points lie inside the central triangle. When their system points are to the left or to the right of the central triangle, they are fully dissociated and therefore act as strong acids or bases respectively.

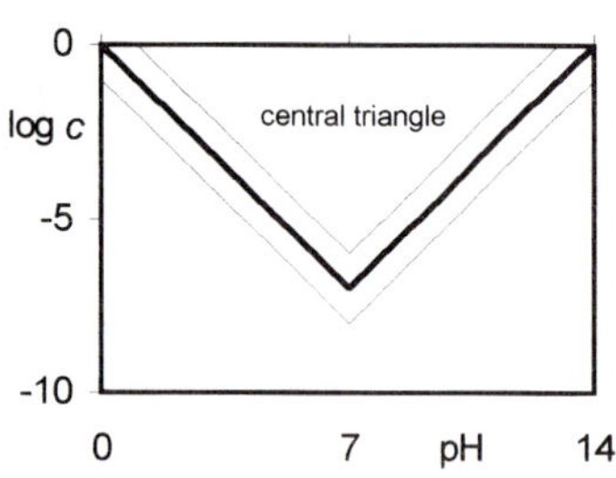

Fig. 2.12 System points well inside or well outside the central triangle lead to simple pH calculations; while those inside the border areas (here indicated by thin lines one unit above and below the border lines) typically can benefit from refinement.

2.3 Refinements

In the examples considered so far we have tacitly assumed that we can use the linear asymptotes rather than the actual, curved lines in the diagram; the examples were selected so that this was mostly the case. In fact, one can simplify the logarithmic concentration diagram somewhat by ignoring the curvature near the system points, and by extending the linear asymptotes all the way to those system points. The resulting *stick diagrams* are easy to sketch. Such stick diagrams were already used by Niels Bjerrum in 1915, in the essay in which he developed the now classical theory of acid–base titrations.

The above examples illustrate the fact that the result of the calculation depends on the location of the system point with respect to the straight lines for $[H^+]$ and $[OH^-]$. More precisely, we can define a *central triangle* in the logarithmic concentration diagram as the area in the diagram bounded at the low concentration side by the lines for $[H^+]$ and $[OH^-]$, as in Fig. 2.10. (The upper bound is given by the solubility of the species considered, for many species in the range between 0.1 and 10 M, and need not concern us here.) Whenever the system point is well within that central triangle, the pH of the acid is given by $\text{pH} \approx (\text{p}C + \text{p}K_a)/2$, and the pH of the corresponding base by $\text{pH} \approx (14 - \text{p}C + \text{p}K_a)/2$. When the system point lies to the left of the line for $[H^+]$ but $\log C$ is larger than -7, the acid behaves as a strong one, and $\text{pH} \approx \text{p}C$. Likewise, when the system point lies to the right of the line for $[OH^-]$ while $\log C > -7$, the base acts as a strong one, and $\text{pH} \approx 14 - \text{p}C$. Finally, for $\log C < -7$, the acid or base is too dilute to affect the pH, which is therefore 7 in the absence of other pH-determining species. These four regions are indicated in Fig. 2.11.

The above examples were picked carefully to illustrate the simplest possible cases. Often, the diagrams are not quite as simple to read, in which case the pH estimate may require some *refinement*. There are two situations in which refinement may be needed: when two or more system points involved in a calculation lie in the border region near the edges of the central triangle, and/or when one or more system points involved in such a calculation lie close together. The former situation is illustrated in Fig. 2.12, where the border region is indicated schematically as extending about one logarithmic unit in pH and log c, and will be considered in more detail below. The latter situation, where the system points lie close together, will be addressed later in this section.

In order to illustrate the types of refinement that may be needed whenever system points lie near the border of the central triangle we will consider three cases: the pH of C M acetic acid, where we will change C by factors of ten, the pH of 1 mM of a hypothetical set of acids HA with pK_as that differ by factors of ten, and finally the pH of the corresponding sodium salts. The only computations needed will involve taking square roots, logarithms, and antilogarithms (i.e., calculating non-integer powers of ten) and can therefore be made with the aid of a scientific pocket calculator.

Figure 2.13 shows a logarithmic concentration diagram modified to show only the two lines representing $[H^+]$ and $\{[A^-]+[OH^-]\}$, i.e., the lines representing the proton gainers and losers for the solution of a single monoprotic acid in water. The lines for $\log\{[A^-]+[OH^-]\}$ are shown for pC values ranging from 0 to 9 with steps of 1. Also shown are the corresponding system points.

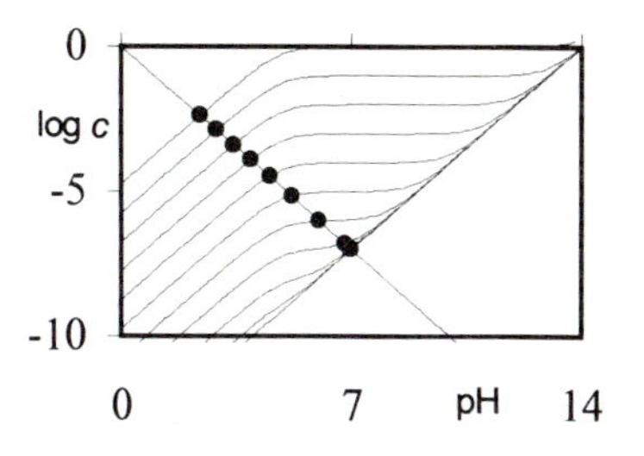

Fig. 2.13 A simplified logarithmic concentration diagram of C M acetic acid (pK_a = 4.76) for log C = 0, −1, −2, ... , −10, showing only the lines for $\log[H^+]$ and $\log\{[A^-]+[OH^-]\}$, with their intersections as solid circles. The corresponding data are listed in Table 2.1.

In each case, the pH is given by the intersection of the line for $\log[H^+]$ (which is independent of the value of pC) and that for $\log\{[A^-]+[OH^-]\}$. As can be seen in Fig. 2.13, the approximation pH = (pC+pK_a)/2 works well for pC = 0, 1, and 2, but becomes questionable at pC = 3, and certainly unreliable at pC-values of 4 or 5, where the curvature in $\log\{[A^-]+[OH^-]\}$ is quite noticeable. For pC = 6 we can use pH = pC, but that simple result will not quite apply to pC = 5 or pC = 7. Here, then, we will indicate how to refine the diagram-based estimates.

Consider the case where pC = 4, where the diagram indicates that the pH will be somewhere in the range between 3 and 5. In that region we can still make the approximation $\{[A^-]+[OH^-]\} \approx [A^-]$ because $[A^-] \gg [OH^-]$. However, we can no longer approximate $[A^-]$ by a simple inverse proportionality to $[H^+]$, but instead we must now use Eqns (1.20) and (1.23), i.e.,

$$[A^-] = C\alpha_0 = \frac{CK_a}{[H^+]+K_a} \tag{2.1}$$

When we combine this with the proton condition $[H^+] = \{[A^-]+[OH^-]\} \approx [A^-]$ it leads to

$$[H^+] \approx [A^-] = \frac{CK_a}{[H^+]+K_a}$$

or

$$[H^+]^2 + [H^+]K_a - CK_a = 0 \tag{2.2}$$

which has the solution

$$[H^+] = \frac{-K_a + \sqrt{K_a^2 + 4CK_a}}{2} \tag{2.3}$$

where the sign in front of the square root must be + because we would otherwise obtain a physically non-realizable negative proton concentration. For $K_a = 10^{-4.76} = 1.74 \times 10^{-5}$ M and $C = 10^{-4}$ M we then find $[H^+] = 3.39 \times 10^{-5}$ M or pH = 4.47. Had we ignored the curvature of the line for $[A^-]$ and instead used pH = (pC+pK_a)/2, we would have found the value 4.37, off by 0.10 from the correct answer.

For $C = 10^{-5}$ M we calculate $[H^+] = 7.10 \times 10^{-6}$ M or pH = 5.15. For $C = 10^{-3}$ M we find $[H^+] = 1.23 \times 10^{-4}$ M or pH = 3.91, which differs by 0.03 from the result estimated from pH = (pC+pK_a)/2 = 3.88. For $C = 10^{-2}$ M we obtain $[H^+] = 4.08 \times 10^{-4}$ M or pH = 3.39, whereas (pC+pK_a)/2 = 3.38. For C = 0.1 M, $[H^+] = 1.31 \times 10^{-3}$ M or pH = 2.88, the same as calculated before using pH = (pC+pK_a)/2.

C = 1 M	pH = 2.38
$C = 10^{-1}$ M	pH = 2.88
$C = 10^{-2}$ M	pH = 3.39
$C = 10^{-3}$ M	pH = 3.91
$C = 10^{-4}$ M	pH = 4.47
$C = 10^{-5}$ M	pH = 5.15
$C = 10^{-6}$ M	pH = 6.01
$C = 10^{-7}$ M	pH = 6.79
$C = 10^{-8}$ M	pH = 6.97
$C = 10^{-9}$ M	pH = 7.00

Table 2.1 The pH of acetic acid (pK_a = 4.76) for various values of its total analytical concentration C.

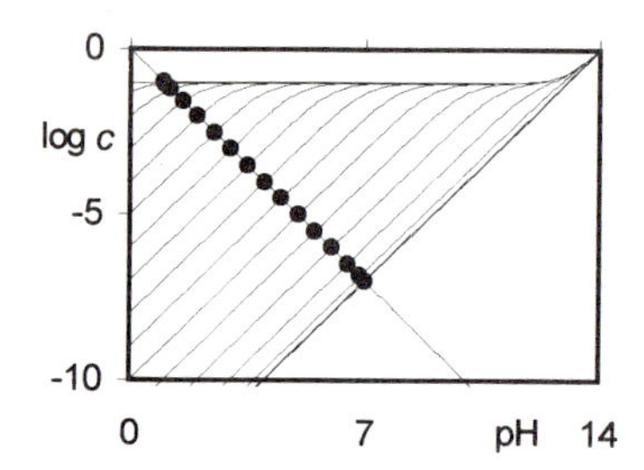

Fig. 2.14 A simplified logarithmic concentration diagram for a 0.1 M solution of a single, monoprotic acid HA for pK_a = –1, 0, 1, 2, ... , 14, 15, showing only the lines for $\log[H^+]$ and $\log\{[A^-]+[OH^-]\}$, with their intersections as solid circles. The corresponding data are listed in Table 2.2.

As we decrease the value of C, the approximation $\{[A^-]+[OH^-]\} \approx [A^-]$ becomes increasingly uncertain. In fact, for $pC \geq 9$, the appropriate approximation would obviously be $\log\{[A^-]+[OH^-]\} \approx \log[OH^-]$, in which case the proton condition reads $[H^+] = [OH^-]$ and therefore yields pH = 7. For $C = 10^{-8}$ M we equate $[A^-]$ approximately to C, so that $[A^-] \approx 10^{-8}$ M, for which we find for the proton condition

$$[H^+] = C + \frac{K_W}{[H^+]}$$

or

$$[H^+]^2 - [H^+]C - K_W = 0 \tag{2.4}$$

$$[H^+] = \frac{C + \sqrt{C^2 + 4K_W}}{2} \tag{2.5}$$

from which we calculate, for $C = 10^{-8}$ M, $[H^+] = 1.05 \times 10^{-7}$ M or pH = 6.98. The same approximation applied to $C = 10^{-7}$ M leads to $[H^+] = 1.62 \times 10^{-7}$ M or pH = 6.79, while $C = 10^{-6}$ M yields $[H^+] = 1.01 \times 10^{-6}$ M or pH = 6.00. Table 2.1 summarizes these results.

In the above example we have used the logarithmic concentration diagram merely to *guide* us to the most appropriate approximation. No specific equations for calculating the pH need to be memorized; instead, the results are obtained directly from the proton condition (an exact result that applies regardless of the particular values of C and pK_a), and the equally general expressions for the concentration fractions α.

We now consider a 0.1 M solution of an acid with pK_a = –1, 0, 1, 2, ..., 13, 14, 15, i.e., ranging all the way from a strong acid to an extremely weak one. Again we illustrate the calculation with a logarithmic concentration diagram in which we display the two sides of the proton condition, $\log[H^+]$ and $\log\{[A^-]+[OH^-]\}$, and their intersection, see Fig. 2.14.

For $pK_a = -1$ we use the diagram to guide us to the approximation $\{[A^-] + [OH^-]\} \approx [A^-] \approx C = 0.1$ M, so that pH = 1.0. This result is not surprising for a strong monoprotic acid. For $pK_a = 5$ through $pK_a = 11$ the system point lies well within the central triangle, and we can again obtain a simple result: the pH lies halfway between pH = pC and pH = pK_a, so that pH = $(pC+pK_a)/2$, which yields a pH of 3.00, 3.50, 4.00, 4.50, 5.00, 5.50, and 6.00 respectively. For $pC = 15$ the diagram shows that $\{[A^-]+[OH^-]\} \approx [OH^-]$ so that the proton condition yields $[H^+] = \{[A^-]+[OH^-]\} \approx [OH^-]$ or pH = 7.00; obviously the acid is now too weak to affect the pH. Refinement is clearly required for pK_a = 0, 1, and 2, as well as for pK_a = 12, 13, and 14.

For pK_a = 0 through 4 we are fully justified in assuming that $\{[A^-]+[OH^-]\} \approx [A^-]$, but we need to take into account the curvature of the line representing $[A^-]$. We already encountered this situation in Eqns (2.1) through (2.3), and we therefore use Eqn (2.3) to find the pH as 1.04, 1.21, 1.57, 2.02, and 2.51 respectively.

pK_a	pH
$pK_a = -1$	pH = 1.00
$pK_a = 0$	pH = 1.04
$pK_a = 1$	pH = 1.21
$pK_a = 2$	pH = 1.57
$pK_a = 3$	pH = 2.02
$pK_a = 4$	pH = 2.51
$pK_a = 5$	pH = 3.00
$pK_a = 6$	pH = 3.50
$pK_a = 7$	pH = 4.00
$pK_a = 8$	pH = 4.50
$pK_a = 9$	pH = 5.00
$pK_a = 10$	pH = 5.50
$pK_a = 11$	pH = 6.00
$pK_a = 12$	pH = 6.48
$pK_a = 13$	pH = 6.85
$pK_a = 14$	pH = 6.98
$pK_a = 15$	pH = 7.00

Table 2.2 The pH of a 0.1 M aqueous solution of a single monoprotic acid for various values of its pK_a.

For pK_a = 12, 13, and 14 the refinement must recognize that $[A^-]$ and $[OH^-]$ are of comparable magnitudes, so that we must construct an auxiliary line representing $\{[A^-] + [OH^-]\}$. In all these cases, the line representing $[A^-]$ runs parallel to that of $[OH^-]$. In a double-logarithmic representation, parallel lines imply that the *ratio* $[A^-] / [OH^-]$ must be constant, and we will now use this. For pK_a = 12 the line for $[A^-]$ runs exactly one logarithmic unit above that for $[OH^-]$, so that $[OH^-] / [A^-] = 0.1$ or $\{[A^-]+[OH^-]\} = (1+0.1) [A^-] = 1.1 [A^-]$. We now construct an auxiliary line segment for $\log\{[A^-]+[OH^-]\}$, which must lie $\log(1.1) = 0.04$ logarithmic units above that for $\log[A^-]$. Its intersection with the line representing $[H^+]$ then occurs at a pH $0.04 / 2 = 0.02$ lower than that for the intersection between $[H^+]$ and $[A^-]$. Since the latter is at pH = (1.00+12.00)/2 = 6.50, the pH is found at 6.50 – 0.02 = 6.48.

For pK_a = 13 the two lines, for $[A^-]$ and $[OH^-]$, coincide in the pH region of interest, around pH 7. Consequently we have $[A^-] = [OH^-]$ so that $\{[A^-]+[OH^-]\} = 2 [A^-]$. In this case, the auxiliary line segment for $\{[A^-]+[OH^-]\}$ must be drawn at a distance $\log(2) = 0.30$ above that for $[A^-]$. Consequently the intersection lies at a pH that is 0.30/2 = 0.15 smaller than 7, or pH = 6.85.

For pK_a = 14 the line for $[OH^-]$ dominates, and we have $[A^-] / [OH^-] = 0.1$ or $\{[A^-] + [OH^-]\} = (1+0.1) [OH^-] = 1.1 [OH^-]$. The auxiliary line segment now runs $\log(1.1) = 0.04$ above the line for $[OH^-]$, so that the pH is found as 7.00 – 0.04/2 = 6.98.

We readily verify that the refinements at pK_a = 11 and 15 are negligble. For pK_a = 11 the line for $[A^-]$ lies 2 logarithmic units above that for $[OH^-]$, so that $\{[A^-]+[OH^-]\} = 1.01 [A^-]$ and $\log\{[A^-]+[OH^-]\} = \log[A^-]+\log(1.01) = \log[A^-]+0.004$, which leads to a correction smaller than ±0.005. Likewise we have a correction of only 0.004 for pK_a = 15.

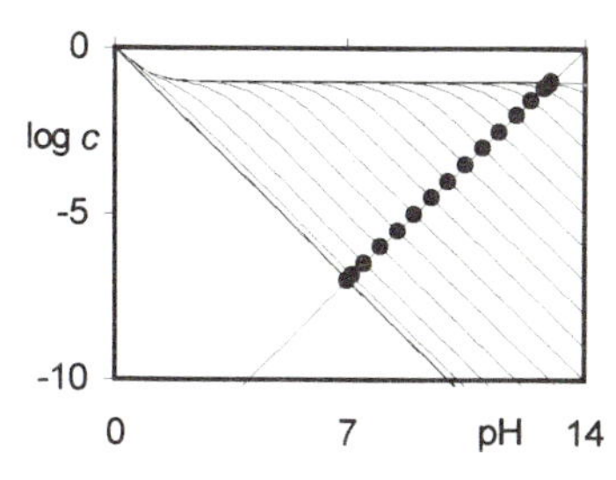

Fig. 2.15 A simplified logarithmic concentration diagram for a 0.1 M solution of the salt of a monoprotic strong base and a monoprotic acid with a pK_a of –1, 0, 1, 2, ... , 14, 15, showing only lines for $\log\{[H^+]+[HA]\}$ and $\log[OH^-]$, with their intersections as solid circles. The corresponding data are listed in Table 2.3.

Finally we consider the pH values of the corresponding salts with a strong monoprotic base, again at a total analytical concentration of 0.1 M. Now the proton condition is $[H^+] + [HA] = [OH^-]$, see Eqn (1.48). In Fig. 2.15 we draw the corresponding logarithmic concentration diagram, again showing only the two terms $\log\{[H^+]+[HA]\}$ and $\log[OH^-]$, and their intersections, for pK_a values ranging from –1 to 15. As in the preceding example, simple results are obtained directly from the diagram for pK_a = 3 through 10, and for pK_a = –1 and pK_a = 15 as well. Again, refinements are necessary where the system point lies close to one of the boundaries of the central triangle, i.e., for pK_a = 0, 1, and 2, and likewise for pK_a = 11 through 14.

First the simple results: for pK_a = –1 we can ignore the contribution of [HA] with respect to that of $[H^+]$, and we therefore find pH = 7.00. For pK_a from 3 to 10 we can, instead, neglect the term $[H^+]$ in $\{[H^+]+[HA]\}$, so that we obtain pH = $(pK_a+14-pC) / 2$ or pH = 8.00, 8.50, 9.00, 9.50, 10.00, 10.50, 11.50, and 11.50 respectively. For a pK_a of 15 we find instead that $[HA] \approx C$ so that pH = 14 – pC = 13.00.

Refinement around pK_a = 1 must use auxiliary lines since $[H^+]$ and [HA] are there of comparable magnitudes. At pK_a = 0 we have $[HA] = 0.1[H^+]$ so that $\{[H^+]+[HA]\} = 1.1[H^+]$. Consequently, the auxiliary line segment for

pK_a	pH
pK_a = –1	pH = 7.00
pK_a = 0	pH = 7.02
pK_a = 1	pH = 7.15
pK_a = 2	pH = 7.52
pK_a = 3	pH = 8.00
pK_a = 4	pH = 8.50
pK_a = 5	pH = 9.00
pK_a = 6	pH = 9.50
pK_a = 7	pH = 10.00
pK_a = 8	pH = 10.50
pK_a = 9	pH = 11.00
pK_a = 10	pH = 11.50
pK_a = 11	pH = 11.98
pK_a = 12	pH = 12.43
pK_a = 13	pH = 12.79
pK_a = 14	pH = 12.96
pK_a = 15	pH = 13.00

Table 2.3 The pH of a 0.1 M aqueous solution of a single monoprotic acid for various values of its pK_a.

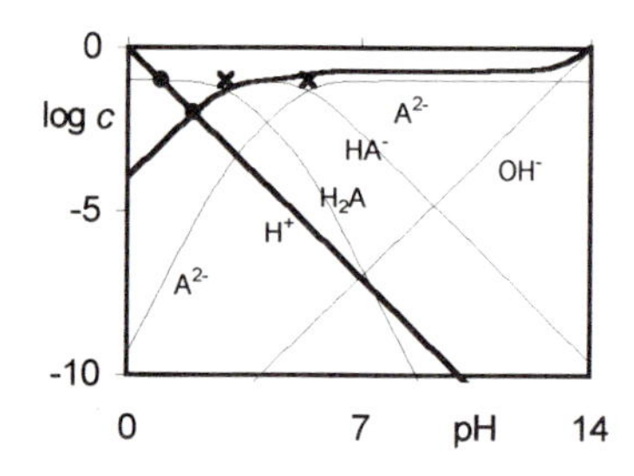

Fig. 2.16 The logarithmic concentration diagram of 0.1 M phthalic acid (pK_{a1} = 2.95, pK_{a2} = 5.41). The sums of the proton gainers and losers are shown as heavy lines. System points are indicated with crosses, the pH-value of phthalic acid as a solid circle, and the auxiliary point at pH = pC as an open circle.

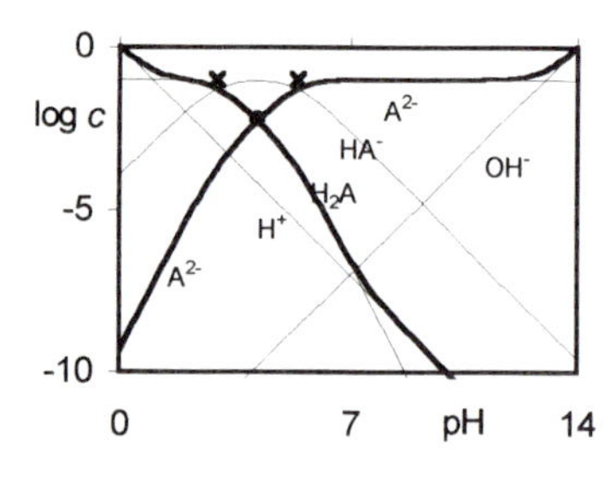

Fig. 2.17 The corresponding diagram of 0.1 M monosodium hydrogen phthalate. Again the proton gainers and losers are shown, as well as the system points, and the pH of the salt.

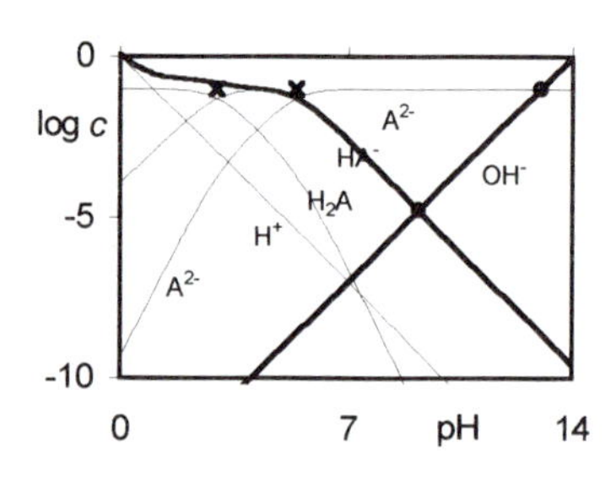

Fig. 2.18 The corresponding diagram of 0.1 M disodium phthalate.

{$[H^+]$+[HA]} runs log(1.1) or 0.04 units above the line for $[H^+]$, and the intersection between the lines for {$[H^+]$+[HA]} and $[OH^-]$ is found at pH = 7.00 + 0.04/2 = 7.02.

For pK_a = 1 we find [HA] = $[H^+]$ so that {$[H^+]$+[HA]} = 2$[H^+]$, which makes the auxiliary line segment lie 0.30 above $[H^+]$, and the intersection therefore 0.30/2 = 0.15 to the right of pH 7, i.e., at pH = 7.15.

For pK_a = 2 we have $[H^+]$ = 0.1[HA], {$[H^+]$+[HA]} = 1.1[HA], and pH = (pK_a+14–pC) / 2 + 0.04 / 2 = 7.52.

Around pK_a = 13 we take into account the curvature of the representation of [HA]. In that case, the proton condition $[H^+]$ + [HA] = $[OH^-]$ reduces to [HA] = $[OH^-]$, and we use eqns (1.19) and (1.22) to find [HA] = $C\alpha_1$ = $[H^+]C$ / ($[H^+]$+K_a). The simplified proton condition [HA] = $[OH^-]$ then yields

$$\frac{[H^+]C}{[H^+]+K_a} = \frac{K_w}{[H^+]} = 0$$

or

$$[H^+]^2C - [H^+]K_w - K_aK_w = 0 \qquad (2.6)$$

$$[H^+] = \frac{-K_w + \sqrt{K_w^2 + 4CK_aK_w}}{2C} \qquad (2.7)$$

Consequently we find pH = 11.98 for pK_a = 11, pH = 12.43 for pK_a = 12, pH = 12.79 for pK_a = 13, and pH = 12.96 for pK_a = 14. Table 2.3 shows the above results.

The above examples illustrate that, for simple cases, we can usually find either simple expressions, or relatively uncomplicated refinements, to compute the pH to within about 0.01. There is seldom a good scientific rationale for making pH measurements to much better than ±0.01.

2.4 Diprotic and triprotic acids, bases, and their salts

Figures 2.16 through 2.20 illustrate the logarithmic concentration diagrams of aqueous solutions of phthalic acid (benzene-1,2-dicarboxylic acid, here abbreviated H_2A) and some of its salts, each at a total analytical concentration C of 0.1 M. For the acid we have the proton condition $[H^+]$ = $[HA^-]$ + 2$[A^{2-}]$ + $[OH^-]$, as illustrated in Fig. 2.16 with heavy lines. In the region where the two heavy lines intersect, the proton condition reduces to $[H^+] \approx [HA^-]$. We are sufficiently far from the region of curvature in the line for HA^-, so that we can use pH = (1+2.95)/2 = 1.98.

For a 0.1 M solution of the acid salt with a strong base, such as potassium hydrogen phthalate, the proton condition reads $[H^+]$ + $[H_2A]$ = $[A^{2-}]$ + $[OH^-]$ or $[H_2A] \approx [A^{2-}]$, see Fig. 2.17, and we find pH = (2.95+5.41)/2 = 4.18.

For the dipotassium salt we have $[H^+]$ + 2$[H_2A]$ + $[HA^-]$ = $[A^{2-}]$ + $[OH^-]$ or $[HA^-] \approx [OH^-]$, see Fig. 2.18, with pH = (5.41 + 13) / 2 = 9.20.

For 0.1 M ammonium hydrogen phthalate we add the ammonium / ammonia couple to the graph, and also include it in the proton condition, which then will read $[H^+] + [H_2A] = [A^{2-}] + [NH_3] + [OH^-]$, as illustrated in Fig. 2.19. Considering the dominant terms on either side yields $[H_2A] \approx [A^{2-}]$, and therefore leads to essentially the same pH as that of the potassium hydrogen salt, $(2.95 + 5.41) / 2 = 4.18$.

For 0.1 M diammonium phthalate, the system point for ammonium now lies at $\log c = -0.7$, because we get two ammonium ions for each phthalate. In this case the proton condition is $[H^+] + 2[H_2A] + [HA^-] = [A^{2-}] + [NH_3] + [OH^-]$ or $[HA^-] \approx [NH_3]$, see Fig. 2.20. Because the two system points are now no longer at the same height in the diagram, we add an auxiliary point (open circle) 0.3 units below and to the left of the system point for ammonia. The pH then follows as the average between this auxiliary point and the second system point for phthalic acid, i.e., $\text{pH} = (5.41 + 9.24 - 0.30) / 2 = 7.18$.

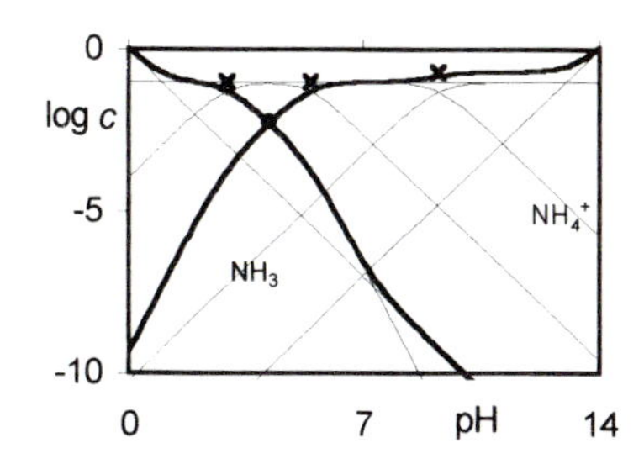

Fig. 2.19 The logarithmic concentration diagram of 0.1 M ammonium hydrogen phthalate. The sums of the proton gainers and losers are shown as heavy lines.

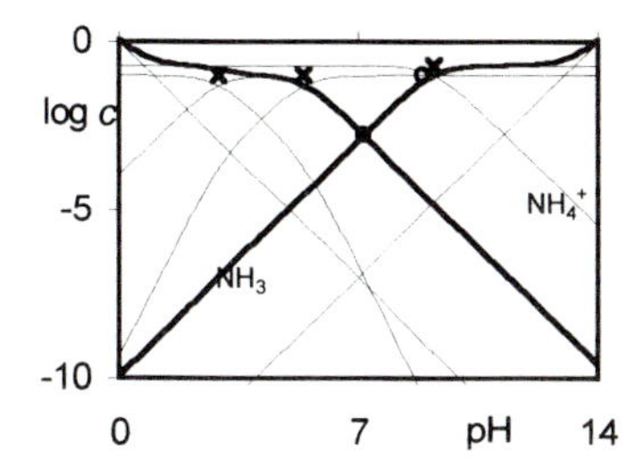

Fig. 2.20 The corresponding diagram of 0.1 M diammonium phthalate. Again the proton gainers and losers are shown.

Figure 2.21 illustrates the logarithmic concentration diagram of phosphoric acid, H_3PO_4, and its various salts, at a total analytical concentration C of 0.1 M. From left to right in Fig. 2.21, the solid points show the intersections of the two parts of the proton condition for 0.1 M H_3PO_4 (with $[H^+] = [H_2PO_4^-] + 2[HPO_4^{2-}] + 3[PO_4^{3-}] + [OH^-]$ or $[H^+] \approx [H_2PO_4^-]$; for 0.1 M NaH_2PO_4 (with $[H^+] + [H_3PO_4] = [HPO_4^{2-}] + 2[PO_4^{3-}] + [OH^-]$ or $[H_3PO_4] \approx [HPO_4^{2-}]$; for 0.1 M Na_2HPO_4 (with $[H^+] + 2[H_3PO_4] + [H_2PO_4^-] = [PO_4^{3-}] + [OH^-]$ or $[H_2PO_4^-] \approx [PO_4^{3-}]$; and for 0.1 M NaH_2PO_4 (with $[H^+] + 3[H_3PO_4] + 2[H_2PO_4^-] + [HPO_4^{2-}] = [OH^-]$ or $[HPO_4^{2-}] \approx [OH^-]$. Note the coefficients: the dissociation of H_3PO_4 to PO_4^{3-} removes three protons, hence the coefficient 3, whereas the formation of PO_4^{3-} from $H_2PO_4^-$ and HPO_4^{2-} involves two or one protons respectively, as reflected in their proton conditions.

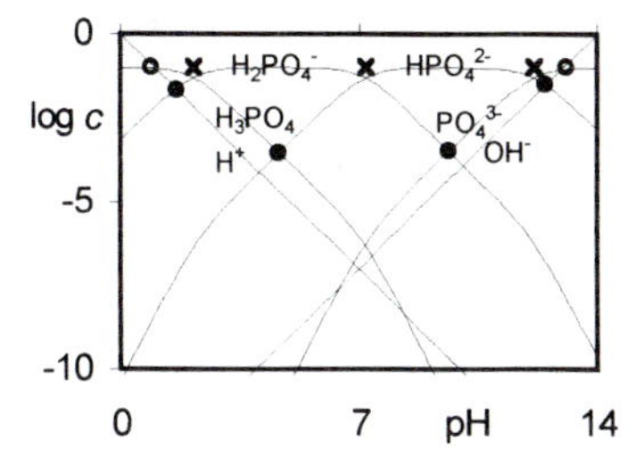

Fig. 2.21 The logarithmic concentration diagram of 0.1 M phosphoric acid ($pK_{a1} = 2.15$, $pK_{a2} = 7.2$, $pK_{a3} = 12.15$) and its sodium salts. The pH-values of H_3PO_4, NaH_2PO_4, Na_2HPO_4, and Na_3PO_4 are shown as solid circles.

Figure 2.22 shows the logarithmic concentration diagram of an aqueous 0.1 M solution of $(NH_4)_2HPO_4$. The proton condition is $[H^+] + 2[H_3PO_4] + [H_2PO_4^-] = [PO_4^{3-}] + [NH_3] + [OH^-]$. Note that the total analytical concentration of ammonium will be 0.2 M, so that the system point for ammonium must be placed at $\log(c) = \log(0.2) = -0.7$ while those for phosphoric acid lie at $\log(c) = -1$, because one mole of $(NH_4)_2HPO_4$ contains two moles of NH_4^+.

We now sketch in the lines for the two sides of the proton condition, i.e., for $[H^+] + 2[H_3PO_4] + [H_2PO_4^-]$ and for $[PO_4^{3-}] + [NH_3] + [OH^-]$, and find that the pH will be around 8. This allows us to simplify the proton condition to $[H_2PO_4^-] \approx 2[NH_3]$. Were we to ignore that the two system points are not at the same height, we would find $\text{pH} = (7.20 + 9.24) / 2 = 8.22$. But the system point for ammonia lies $\log(2) = 0.30$ higher. We therefore place an auxiliary point 0.3 lower, and to the left, of the system point, so that it lies at the same height as the system points for phosphate (shown as an open circle in Fig. 2.22). We then find the pH at $(7.20 + 8.94) / 2 = 8.07$.

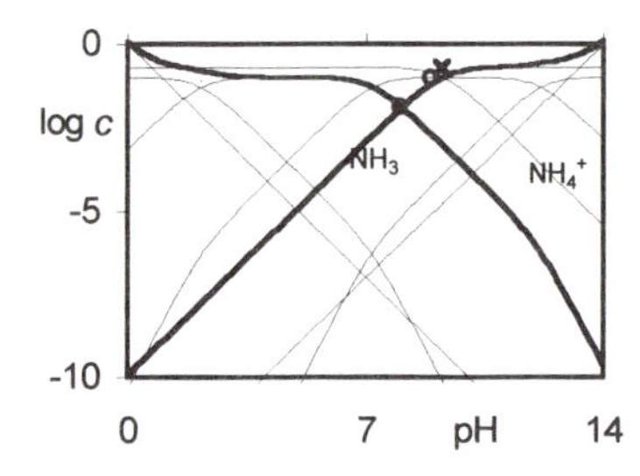

Fig. 2.22 The corresponding diagram of 1 mM $(NH_4)_2HPO_4$ is further complicated by the presence of the lines for NH_4^+ and NH_3. The heavy lines show $\{[H+] + 2[H_3PO_4] + [H_2PO_4^-]\}$ and $\{[PO_4^{3-}] + [NH_3] + [OH^-]\}$ respectively.

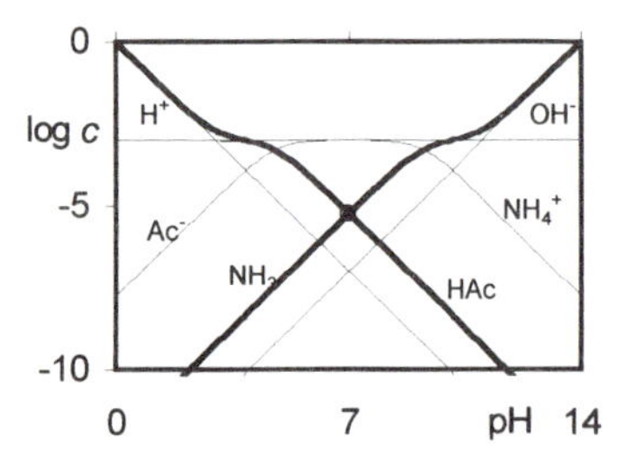

Fig. 2.23 The logarithmic concentration diagram of 1 mM ammonium acetate, with its proton condition, $[H^+] + [HAc] = [NH_3] + [OH^-]$.

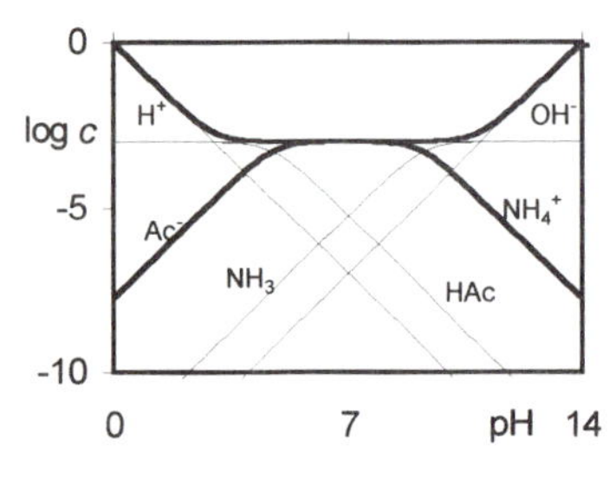

Fig. 2.24 The same diagram but with another, mathematically equivalent but computationally useless form of the proton condition, $[H^+] + [NH_4^+] = [Ac^-] + [OH^-]$.

2.5 More complicated cases

The methods used so far rely on the proton condition. This is fine as long as we realize that it is possible to find equations that are *mathematically* equivalent to the proton condition, but can be quite different *computationally*. We will illustrate this with a case already discussed in connection with Fig. 2.9, the pH of 1 mM ammonium acetate, for which the mass balance equations read $[NH_4^+] + [NH_3] = C$ and $[HAc] + [Ac^-] = C$, while the charge balance relation is $[H^+] + [NH_4^+] = [Ac^-] + [OH^-]$. Earlier we combined these to the proton condition $[H^+] + [HAc] = [NH_3] + [OH^-]$, and then found the pH at the intersection of the curves for $\log\{[H^+]+[HAc]\}$ and $\log\{[NH_3]+[OH^-]\}$, see Fig. 2.23. If we were to use instead the charge balance equation $[H^+] + [NH_4^+] = [Ac^-] + [OH^-]$ we would run into a problem: in that case the intersection of $\log\{[H^+]+[NH_3]\}$ and $\log\{[HAc]+[OH^-]\}$ occurs under far too sharp an angle to be useful, see Fig. 2.24. The rule that the proton condition should not contain the starting species, in this case $[NH_4^+]$ and $[Ac^-]$, avoids this problem.

However, in mixtures, such a rule cannot always be applied, in which case the precise form of the proton condition may not be unique. The resulting ambiguity is seldom a problem, because when we run into such a problem we can always find an alternative form of the proton condition that can be used. The problem is discussed here only to alert the reader that, occasionally, some algebraic reorganization of the proton condition may be required before it will yield its secret.

To illustrate this, consider a solution which is C_1 M in acetic acid and C_2 M in sodium acetate. In that case we have the mass balance equations $[HAc] + [Ac^-] = C_1 + C_2$, and $[Na^+] = C_2$, together with the charge balance relation $[H^+] + [Na^+] = [Ac^-] + [OH^-]$. In this case it is not possible to exclude simultaneously the terms [HAc], $[Na^+]$, and $[Ac^-]$ associated with the starting species HAc and NaAc. Substitution of C_2 for $[Na^+]$ yields $[H^+] + [Na^+] = [Ac^-] + [OH^-]$, while further substitution of $[HAc] + [Ac^-] - C_1$ for C_2 leads to $[H^+] + [HAc] = C_1 + [OH^-]$. Figures 2.25 and 2.26 illustrate these two proton conditions for $C_1 = C_2 = 10$ mM. In both cases, the pH is readily found after the curvature of one of the two lines is taken into account.

2.6 Iterative methods

While it is often possible to make refinements that result in quadratic expressions, or that can be solved by auxiliary line segments, this is by no means always true, especially when successive pK_a values lie close together, as in, e.g., glutaric acid ($pK_{a1} = 4.34$, $pK_{a2} = 5.43$), citric acid ($pK_{a1} = 3.13$, $pK_{a2} = 4.76$, $pK_{a3} = 6.40$), or ethylene diamine tetraacetic acid ($pK_{a2} = 1.5$, $pK_{a3} = 2.0$, $pK_{a4} = 2.68$). In such cases the user must make a choice: either be satisfied with a cruder pH estimate (sometimes justified because the corresponding pK_a-values may not be known to high precision either), or use one of the more sophisticated approaches, such as will be described in the next two sections.

So far, our method has been based on solving the proton condition. In relatively simple cases, a sketched logarithmic concentration diagram will show what approximations are suitable, and with luck such approximations will reduce the problem to one that can be handled with the simple tools of general-purpose scientific pocket calculators. But there is no good reason to stop there: as long as we can formulate the proton condition, which is an explicit polynomial expression in $[H^+]$, the problem is mathematically well-posed, so that an iterative solution can be found.

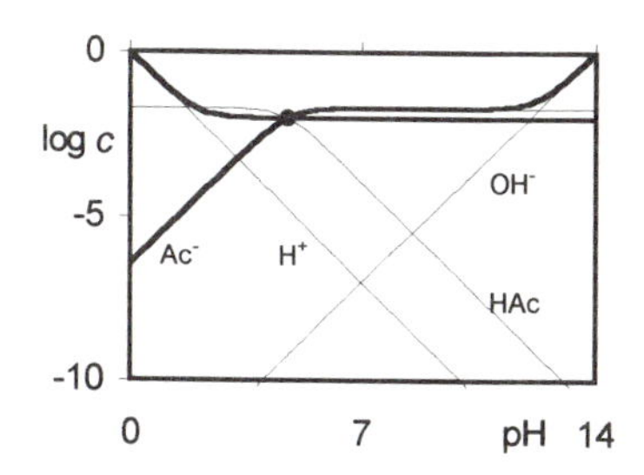

Fig. 2.25 The logarithmic concentration diagram of a solution 10 mM in acetic acid and 10 mM in sodium acetate. The heavy lines represent the two sides of the proton condition $[H^+] + [Na^+] = [Ac^-] + [OH^-]$.

Given the central role of the proton condition in this section, we will now formulate it in a general form. First we look at the proton condition of the single, monoprotic acid HA, which is $[H^+] = [A^-] + [OH^-]$. We substitute $[A^-] = C\alpha_0$ and write the result in standard form (i.e., with the right-hand side equal to zero), so that we obtain $C\alpha_0 - \Delta = 0$, where $\Delta = [H^+] - [OH^-] = [H^+] - K_w/[H^+]$.

Now consider an aqueous solution of H_3PO_4, for which the proton condition reads

$$[H^+] = [H_2PO_4^-] + 2[HPO_4^{2-}] + 3[PO_4^{3-}] + [OH^-] \tag{2.8}$$

Upon substituting $C\alpha_2$, $C\alpha_1$, and $C\alpha_0$ for $[H_2PO_4^-]$, $[HPO_4^{2-}]$, and $[PO_4^{3-}]$ respectively, and rearranging to bring the equation into standard form, we obtain

$$C\,(\alpha_2 + 2\alpha_1 + 3\alpha_0) - \Delta = 0 \tag{2.9}$$

Or take an aqueous solution of the acid salt Na_2HPO_4, for which the proton condition reads

$$[H^+] + 2[H_3PO_4] + [H_2PO_4^-] = [PO_4^{3-}] + [OH^-] \tag{2.10}$$

which can be rearranged to

$$C\,(-2\alpha_3 - \alpha_2 + \alpha_0) - \Delta = 0 \tag{2.11}$$

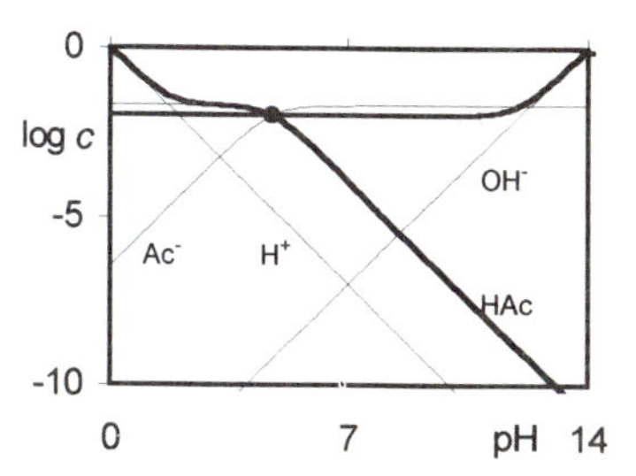

Fig. 2.26 The logarithmic concentration diagram of a solution 10 mM in acetic acid and 10 mM in sodium acetate. The heavy lines represent the two sides of the equivalent proton condition $[H^+] + [HAc] = C_1 + [OH^-]$.

In all these examples we have written the proton condition in the simple form $FC - \Delta = 0$, where F is a function of the various concentration fractions. In this form, the proton condition can readily be generalized to read, for any mixture,

$$\sum_i F_{i,a} C_i - \Delta = 0 \tag{2.12}$$

where F_a is the *acid dissociation function.* For bases it is somewhat more convenient to use the complementary *base dissociation function* F_b, for which the corresponding expression is

$$\sum_i F_{i,b} C_i + \Delta = 0 \tag{2.13}$$

We will encounter both functions in Chapter 3, where they will figure prominently in the theory of acid–base titrations. Comparison of eqns (2.12) and (2.13) shows that $F_b = -F_a$, so that we can use one or the other. Table 2.4 shows the functions F_a and F_b for the same species for which Table 1.1 lists the proton condition.

Table 2.4 The acid dissociation function F_a for *o*-phosphoric acid and some of its salts

H_3PO_4:	$F_a = -F_b = \quad + \alpha_{P,2} + 2\alpha_{P,1} + 3\alpha_{P,0}$
NaH_2PO_4:	$F_a = -F_b = -\alpha_{P,3} + \alpha_{P,1} + 2\alpha_{P,0}$
$NH_4H_2PO_4$:	$F_a = -F_b = -\alpha_{P,3} + \alpha_{P,1} + 2\alpha_{P,0} + \alpha_{A,0}$
Na_2HPO_4:	$F_a = -F_b = -2\alpha_{P,3} - \alpha_{P,2} + \alpha_{P,0}$
$NaNH_4HPO_4$:	$F_a = -F_b = -2\alpha_{P,3} - \alpha_{P,2} + \alpha_{P,0} + \alpha_{A,0}$
$(NH_4)_2HPO_4$:	$F_a = -F_b = -2\alpha_{P,3} - \alpha_{P,2} + \alpha_{P,0} + 2\alpha_{A,0}$
Na_3PO_4:	$F_a = -F_b = -3\alpha_{P,3} - 2\alpha_{P,2} - \alpha_{P,1}$
$Na_2NH_4PO_4$:	$F_a = -F_b = -3\alpha_{P,3} - 2\alpha_{P,2} - \alpha_{P,1} + \alpha_{A,0}$
$Na(NH_4)_2PO_4$:	$F_a = -F_b = -3\alpha_{P,3} - 2\alpha_{P,2} - \alpha_{P,1} + 2\alpha_{A,0}$
$NH_4)_3PO_4$:	$F_a = -F_b = -3\alpha_{P,3} - 2\alpha_{P,2} - \alpha_{P,1} + 3\alpha_{A,0}$

The index P refers to phosphoric acid, H_3PO_4, while the index A refers to ammonium, NH_4^+. Explicit expressions for $\alpha_{P,3}$ through $\alpha_{P,0}$ are given in eqns (1.27) through (1.30), and $\alpha_{A,0}$ is specified in eqn (1.23).

Here we will take one of the species listed in Table 2.4, $(NH_4)_2HPO_4$, and analyze how its acid dissociation function F_a is obtained. Upon dissolving $(NH_4)_2HPO_4$ in water, we start with one mole of HPO_4^{2-} and two moles of NH_4^+ per mole of $(NH_4)_2HPO_4$. Dissociation of one proton then yields one mole of PO_4^{3-} and two moles of NH_3, while association of one proton leads to $H_2PO_4^-$, and association of two protons to H_3PO_4. Hence we write

$$F_a = -\alpha_{H_2PO_4^-} - 2\alpha_{H_3PO_4} + \alpha_{PO_4^{3-}} + 2\alpha_{NH_3}$$

$$= \frac{-[H^+]^2 K_{P1} - 2[H^+]^3 + K_{P1}K_{P2}K_{P3}}{[H^+]^3 + [H^+]^2 K_{P1} + [H^+]K_{P1}K_{P2} + K_{P1}K_{P2}K_{P3}} + \frac{2K_A}{[H^+] + K_A} \tag{2.14}$$

or, in the condensed notation of Table 2.4, $F_a = -\alpha_{P,2} - 2\alpha_{P,3} + \alpha_{P,0} + 2\alpha_{A,0}$. The logic used here also makes clear why F_b is equal to $-F_b$, because the reasoning for F_b would be identical except that we would interchange the signs for the contributions to proton *dis*sociation and *as*sociation respectively.

We saw in section 1.3 that concentration fractions can always be written explicitly in terms of the equilibrium constants and $[H^+]$, and the same therefore applies to eqns (2.12) and (2.13). Since the pH is defined by these expressions of the proton condition, all we have to do to find the pH is to determine the smallest real, positive root for $[H^+]$. Furthermore, because the proton condition written in this form is monotonic in $[H^+]$, standard methods of finding real roots of single-parameter expressions, such as the Newton–Raphson method, often suffice. Such methods are already implemented in many spreadsheets, under such names as SolveFor in QuattroPro, and GoalSeek in Excel.

A problem with this approach is that the default criterion for quitting the iteration may be too loose. For example, when we try to compute the pH of 1 M Na_3PO_4 with a Newton–Raphson algorithm that stops the iteration when the criterion is met to within 10^{-6}, we will not find a reliable answer, since $[H^+]$ in this case is of the order of 10^{-13}. Fortunately, in most commercial Newton–Raphson algorithms the user can set the criterion to a value much smaller than its default setting, and this is often necessary. The computational accuracy of a more general algorithm, such as the multi-parameter non-linear least-squares Solver in Excel, is more easily adjusted, and can therefore be used instead of the Newton–Raphson method.

With all such computations it is useful to have a starting value that is not too far removed from the final answer, in order to keep the search from going astray. An initial estimate can readily be obtained from the logarithmic concentration diagram. It can also be found by calculating F_a or F_b as a function of $[H^+]$ (using steps of, say, factors-of-ten) or pH (e.g., from 0 to 14 in steps of 1). Where the function crosses zero it will change its sign; take the nearest value of $[H^+]$ as your initial guess. Alternatively you might plot F_a or F_b, although that is usually more work than computing a simple 15-point table and looking for the sign change. Even without a close initial estimate, the Newton–Raphson method usually works well for this type of problems, because the functions F_a or F_b are monotonic, i.e., they have neither minima nor maxima, but either keep increasing or decreasing, with increasing values of $[H^+]$.

Alternatively, once we have located the pH interval of the zero crossing to within, say, one pH unit, we might repeat the process within that unit for pH increments of 0.1, and subsequently in the resulting 0.1-wide zero-crossing interval use steps of 0.01 (or interpolate), to arrive directly at the desired result by trial and error. Given the fact that we are seldom interested in the third decimal place of the pH value, even such a simple method works.

Other approaches (such as using hyperbolic functions, see Rang 1976, Herman et al. 1990) have been proposed to deal with the complication that $[H^+]$ can be very small, and therefore requires a correspondingly small criterion for terminating the iteration. For example, Rang (1976) showed that the use of hyperbolic functions avoids the above problem; unfortunately, this method works only for monoprotic acids and bases, see Herman et al. (1990). Fortunately there is seldom a need for such special methods as long as the convergence criterion is set properly.

2.7 Equation solvers

As our last method we will briefly mention general purpose programs that can solve sets of simultaneous equations, using matrix algebra. The required input information typically consists of all species with their valencies and concentrations (from which the program can find the mass and charge balance relations), plus all relevant equilibrium constants. The equilibrium and balance expressions form a well-defined set of simultaneous equations that

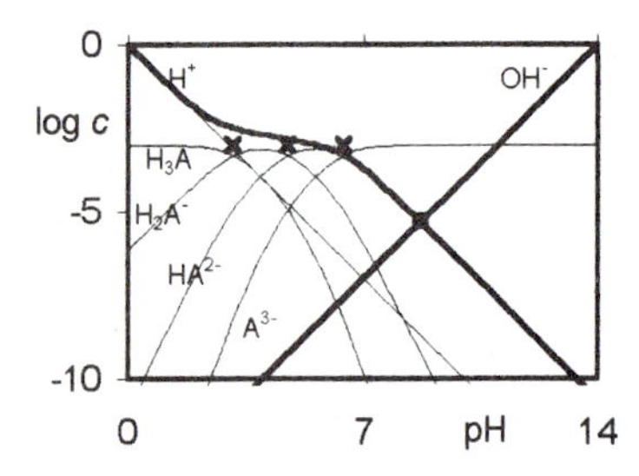

Fig. 2.27 The logarithmic concentration diagram of a solution 1 mM trisodium citrate, and its proton gainers and losers (heavy lines), $\{[H^+] + 3[H_3A] + 2[H_2A^-] + 2[HA^{2-}]\}$ and $[OH^-]$ respectively.

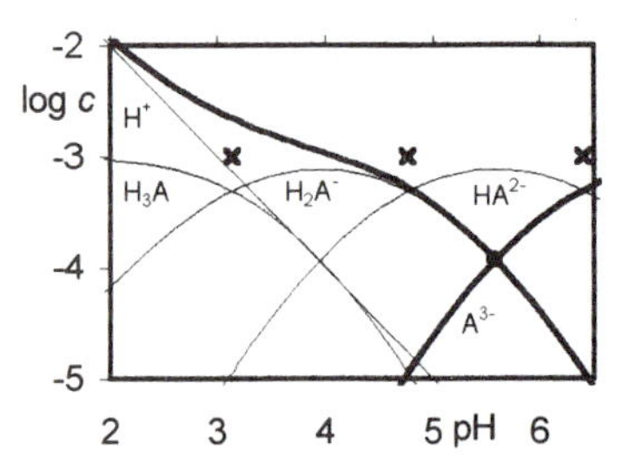

Fig. 2.28 Detail of the logarithmic concentration diagram of a solution 1 mM disodium hydrogen citrate, and its proton gainers and losers (heavy lines), $\{[H^+]+2[H_3A]+[H_2A^-]\}$ and $\{[A_3^-]+[OH^-]\}$ respectively.

pH	[H+]	$F_aC-\Delta$
5.50	3.16E-06	4.72E-05
5.60	2.51E-06	-7.44E-06
5.70	2.00E-06	-6.20E-05
5.58	2.63E-06	3.45E-06
5.59	2.57E-06	-2.00E-06
5.60	2.51E-06	-7.44E-06

Table 2.5 The proton condition for 1 mM Na_2HCit as a function of pH.

can be solved by computer. Such methods are typically less transparent to the user, and require specialized software, but they will do the job, and will find the concentrations of all species involved.

While such a method might be gross overkill if only the pH were required, the approach is readily extendable to systems in which there are more than one central parameter, as with multi-ligand metal complexation systems. They are especially useful when multi-nuclear complexes are formed. The prototypical computer programs in this field were indeed developed by Sillén et al. for such complicated inorganic systems.

2.8 Sample calculations

In order to illustrate the methods explained so far we will here first compute the pH of 1 mM citric acid and its sodium salts. These are still fairly simple systems, and will highlight the various points made so far. Citric acid has three pK_a-values that lie close together (at 3.13, 4.76, and 6.40); a concentration of 1 mM is used because it brings the first system point (at pH = pK_{a1}, pc = pC) close to the edge of the central triangle. For notational compactness we will represent citric acid as H_3A.

We start with trisodium citrate as the simplest of the four. The proton condition is $[H^+]+3[H_3A]+2[H_2A^-]+[HA^{2-}] = [OH^-]$, and the logarithmic concentration diagram of Fig. 2.27 shows that it can be simplified for all practical purposes to $[HA^{2-}] = [OH^-]$. The pH can therefore be found directly as pH = $(pK_{a3}+pK_w-pC)/2 = (6.40+11.00)/2 = 8.70$. No further refinement is necessary.

For the disodium salt Na_2HA, the proton condition reads $[H^+] + 2[H_3A] + [H_2A^-] = [A^{3-}] + [OH^-]$, see Fig. 2.28. A first estimate can be based on $[H_2A^-] \approx [A^{3-}]$; when we then neglect the presence of K_{a1} we find as our first estimate pH $\approx (pK_{a2} + pK_{a3})/2 = (4.76 + 6.40)/2 = 5.58$. Here we will demonstrate the trial-and-error method based on finding where the proton condition in standard form is zero, i.e., where $[H^+] + 2[H_3A] + [H_2A^-] - [A^{3-}] - [OH^-] = 0$. We therefore use a spreadsheet with a built-in Newton–Raphson algorithm, to find the zero crossing of $[H^+] + 2[H_3A] + [H_2A^-] - [A^{3-}] - [OH^-] = [H^+] + \{2[H^+]^3 + [H^+]^2K_{a1} - K_{a1}K_{a2}K_{a3}\} / \{[H^+]^3 + [H^+]^2K_{a1} + [H^+]K_{a1}K_{a2} + K_{a1}K_{a2}K_{a3}\} - K_w/[H^+]$, with $[H^+] = 10^{-5.58} = 2.6 \times 10^{-6}$ as our starting estimate. Provided that we have set the convergence criterion to a sufficiently small value, this will yield $[H^+] = 2.59609 \times 10^{-6}$, at which the proton condition differs from zero by about 6×10^{-11}. The corresponding pH is 5.58635, quite close to the original estimate of 5.58.

Alternatively, in Table 2.5 we compute a few pH-values of $[H^+] + 2[H_3A] + [H_2A^-] - [A^{3-}] - [OH^-] = [H^+] + \{2[H^+]^3 + [H^+]^2K_{a1} - K_{a1}K_{a2}K_{a3}\} / \{[H^+]^3 + [H^+]^2K_{a1} + [H^+]K_{a1}K_{a2} + K_{a1}K_{a2}K_{a3}\} - K_w/[H^+]$ in the neighborhood of pH 5.58. In this way we can quickly home in on the answer, pH = 5.59 or, through linear interpolation, to pH = 5.58 + 3.46 / (3.46 + 2.00) × 0.01 = 5.5863.

The proton condition for the monosodium salt NaH_2A reads $[H^+] + [H_3A] = [HA^{2-}] + 2[A^{3-}] + [OH^-]$, or to a first approximation $[H_3A] \approx [HA^{2-}]$, as illustrated in Fig. 2.29. Neglecting the presence of pK_{a3} then yields as a first estimate $pH \approx (pK_{a1} + pK_{a2})/2 = (3.13 + 4.76)/2 = 3.94$. Again, we reshuffle the terms of the proton condition to bring it into standard form, $[H^+] + [H_3A] - [HA^{2-}] - 2[A^{3-}] - [OH^-] = [H^+] + \{[H^+]^3 - [H^+]K_{a1}K_{a2} - 2K_{a1}K_{a2}K_{a3}\} / \{[H^+]^3 + [H^+]^2K_{a1} + [H^+]K_{a1}K_{a2} + K_{a1}K_{a2}K_{a3}\} - K_w/[H^+]$, and use either a Newton–Raphson method, a non-linear least-squares algorithm such as the Excel Solver, or a short table, to obtain pH = 4.09144. Likewise a short list such as shown in Table 2.6 yields the value 4.09, and linear interpolation can be used to refine this to pH = 4.09 + 1.07 / (1.07+6.34) × 0.01 = 4.0914.

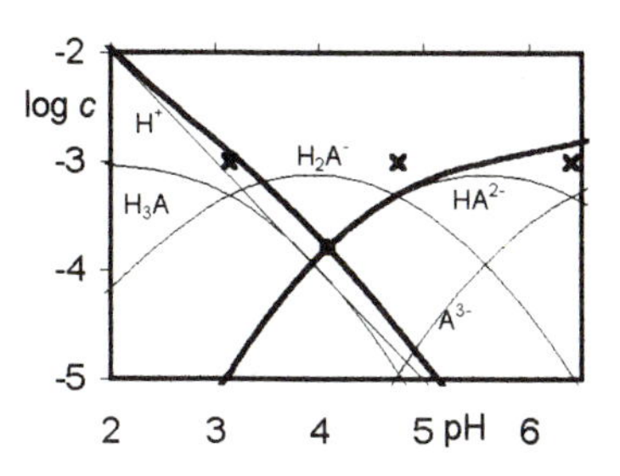

Fig. 2.29 Detail of the logarithmic concentration diagram of a solution 1 mM sodium dihydrogen citrate, and its proton gainers and losers (heavy lines), $\{[H^+]+[H_3A]\}$ and $\{[HA_2^-]+2[A_3^-]+[OH^-]\}$ respectively.

pH	[H+]	$F_aC - \Delta$
4.00	1.00E-04	6.92E-05
4.10	7.94E-05	–6.34E-06
4.20	7.76E-05	–1.37E-05
4.08	8.32E-05	8.50E-06
4.09	8.13E-05	1.07E-06
4.10	7.94E-05	–6.34E-06

Table 2.6 The proton condition for 1 mM NaH_2Cit as a function of pH.

Finally we consider the pH of the acid, H_3A. The proton condition is $[H^+] = [H_2A^-] + 2[HA^{2-}] + 3[A^{3-}] + [OH^-]$, which we simplify for a first approximation to $[H^+] \approx [H_2A^-]$, see Fig. 2.30. When we take into account the curvature of $[H_2A^-]$, but neglect the presence of both pK_{a2} and pK_{a3} in the computation of $[H_2A^-]$, we derive eqns (2.1) through (2.3), which lead to $[H^+] \approx 5.67 \times 10^{-4}$ M or pH ≈ 3.25. In standard form, the proton condition is $[H^+] + \{-[H^+]^2K_{a1} - 2[H^+]K_{a1}K_{a2} - 3K_{a1}K_{a2}K_{a3}\} / \{[H^+]^3 + [H^+]^2K_{a1} + [H^+]K_{a1}K_{a2} + K_{a1}K_{a2}K_{a3}\} - K_w/[H^+]$, for which iteration produces pH = 3.2341 while a short trial-and-error calculation (see Table 2.7) with linear interpolation yields pH = 3.23 + 8.16 / (8.16+11.6) × 0.01 = 3.2341.

As our final example we will indicate (though not quite work out numerically) how one would treat a Britton–Robinson buffer, i.e., a mixture of boric acid (pK_{a1} = 9.24, pK_{a2} = 12.7, pK_{a3} = 13.8), citric acid (pK_{a1} = 3.13, pK_{a2} = 4.76, pK_{a3} = 6.40), diethylbarbituric acid (pK_a = 7.98), and phosphoric acid (pK_{a1} = 2.15, pK_{a2} = 7.20, pK_{a3} = 12.15), to which the user adds NaOH to adjust the pH to a desired value.

We will use the subscripts B, C, D, and P to denote boric, citric, diethylbarbituric, and phosphoric acid respectively, and S for sodium hydroxide. Following the format of eqn (2.12), we then write the proton condition as

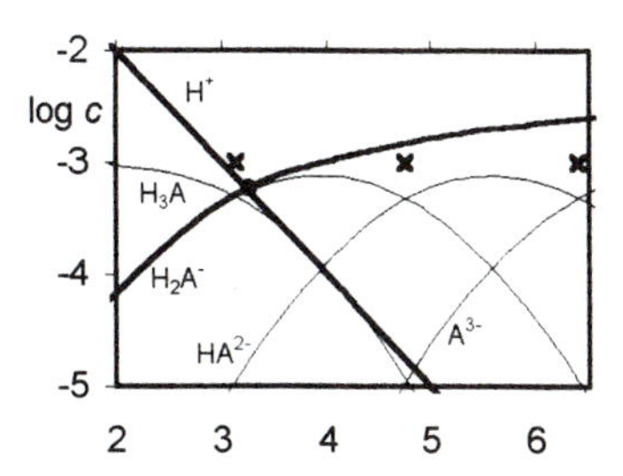

Fig. 2.30 Detail of the logarithmic concentration diagram of a solution 1 mM citric acid (pK_{a1} = 3.13, pK_{a2} = 4.76, pK_{a3} = 6.40) and the proton gainers and losers (heavy lines): $[H^+]$ and $\{[H_2A^-] + 2[HA_2^-] + 3[A_3^-] + [OH^-]\}$.

$$F_BC_B + F_CC_C + F_DC_D + F_PC_P + F_SC_S - \Delta$$

$$= \frac{[H^+]^2K_{B1} + 2[H^+]K_{B1}K_{B2} + 3K_{B1}K_{B2}K_{B3}}{[H^+]^3 + [H^+]^2K_{B1} + [H^+]K_{B1}K_{B2} + K_{B1}K_{B2}K_{B3}}C_B$$

$$+ \frac{[H^+]^2K_{C1} + 2[H^+]K_{C1}K_{C2} + 3K_{C1}K_{C2}K_{C3}}{[H^+]^3 + [H^+]^2K_{C1} + [H^+]K_{C1}K_{C2} + K_{C1}K_{C2}K_{C3}}C_C$$

$$+ \frac{[H^+]^2K_{P1} + 2[H^+]K_{P1}K_{P2} + 3K_{P1}K_{P2}K_{P3}}{[H^+]^3 + [H^+]^2K_{P1} + [H^+]K_{P1}K_{P2} + K_{P1}K_{P2}K_{P3}}C_P$$

$$+ \frac{[H^+]C_D}{[H^+] + K_{D1}} - C_S - [H^+] + \frac{K_w}{[H^+]} \qquad (2.15)$$

pH	[H+]	$F_aC - \Delta$
3.10	7.94E-04	2.96E-04
3.20	6.31E-04	6.93E-05
3.30	5.01E-04	−1.24E-04
3.22	6.03E-04	2.82E-05
3.23	5.89E-04	8.16E-06
3.24	5.75E-04	−1.16E-05

Table 2.7 The proton condition for 1 mM H_3Cit as a function of pH.

where the acid dissociation function F_S is −1 for the strong base NaOH. Despite its length (which reflects the rather complicated nature of this buffer mixture, with five concentrations and eleven equilibrium constants), eqn (2.15) is a straightforward expression with one unknown, $[H^+]$, and its root for given values of the concentrations and equilibrium constants is found just as readily as that for citric acid.

By adding strong base or acid, the pH of such a buffer mixture can be changed. This is, of course, a titration of that buffer mixture. In the next chapter we will see how such a titration can be described quantitatively.

3 Titrations

The primary purpose of a titration is to measure an unknown amount of a substance in a sample through chemical reaction with a known amount of a suitable reagent, in such a way that we can observe when an equivalent amount of reagent has been added to the sample. From a knowledge of that *equivalence point* we can then determine the unknown amount. Typically, a known volume of sample is taken, so that the unknown *amount* is the product of the *known* sample *volume* V_s and the *unknown* sample *concentration* C_s. When the equivalence point is reached after reaction of a volume V_t of titrant of concentration C_t, the two amounts are related to each other through the appropriate *stoichiometric factor* s. For the titration of a monoprotic acid with a monoprotic base the stoichiometric factor s is 1, for the titration of citric acid with NaOH, s will be 3 (but 1 or 2 for the titration with NaOH of the equally triprotic H_3PO_4, depending on which equivalence point we use), for the titration of a monoprotic acid with a strong diprotic base such as calcium hydroxide we have $s = ½$, for the titration of citric acid with $Ca(OH)_2$, $s = 3/2$, and so on. We will come back to the value of s in Section 3.4. In all cases, the unknown concentration can then be calculated as $C_s = sC_tV_t/V_s$.

While this relation only yields the unknown concentration C_s, a more complete analysis shows that a titration can provide additional information about the sample. In the case of acid–base titrations, such additional information consists of the acid dissociation constants of the sample. Indeed, titrations are often used to determine such constants.

3.1 Progress and titration curves

Below we will first consider the titration of a single, strong monoprotic acid with a single, strong monoprotic base. Subsequently we will generalize the approach to a weak acid, to diprotic and polyprotic acids and bases, and finally to the titration of arbitrary mixtures of these. For the sake of simplicity we will, as before, restrict the discussion to aqueous solutions.

The simplest titration is that of a single, strong monoprotic acid with a single, strong monoprotic base. In order to make the discussion more concrete, we will use HCl and NaOH as our model strong acid and strong base respectively. We will place a sample volume V_a containing the *a*cid HCl of (as yet unknown) concentration C_a in a titration vessel. We will fill a buret with a solution containing the *b*ase NaOH at a known concentration C_b. During the titration we gradually add base from the buret to the sample, all the time measuring the total volume V_b dispensed from the buret.

Since material is added to the sample from the outside, we have what is called an *open* system, and we must accordingly modify the mass balance conditions. In the present example this is most readily done by considering the *spectator ions* Cl^- and Na^+, so called because they do not participate in the titration reactions.

Let the original sample have a volume V_a and contain a concentration C_a of HCl. For a strong acid, which is fully dissociated, this implies $[Cl^-] = C_a$. Upon addition of a volume V_b of base NaOH, the total volume in the titration vessel increases from V_a to V_a+V_b. Since the total amount of Cl^- in the vessel remains constant, its concentration is diluted by a factor $V_a/(V_a+V_b)$.

There is a similar dilution of the sodium ions of the titrant upon its addition to the sample: when the titrant has a sodium concentration C_b, the corresponding concentration in the titration vessel will be diluted by a factor $V_b/(V_a+V_b)$. Consequently we have the mass balance equations

$$[Cl^-] = \frac{C_a V_a}{V_a + V_b} \tag{3.1}$$

$$[Na^+] = \frac{C_b V_b}{V_a + V_b} \tag{3.2}$$

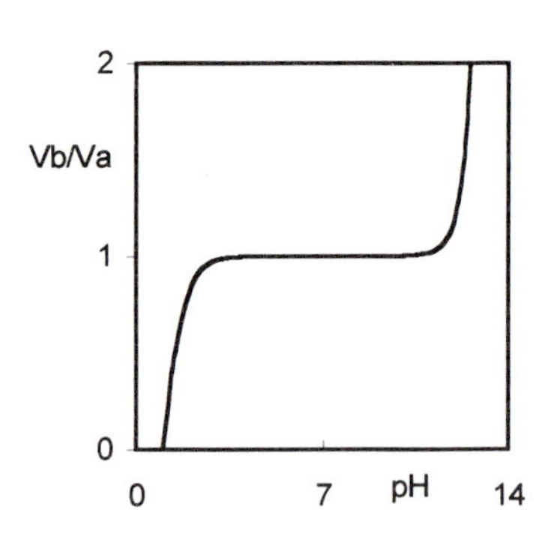

Fig. 3.1a The progress curve for the titration of 0.1 M HCl with 0.1 M NaOH as computed from Eqn. (3.4) with $C_a = C_b = 0.1$. The vertical axis shows V_b/V_a in units of C_a/C_b.

We now consider the charge balance relation. Since the only ions present in the solution being titrated are H^+, Na^+, Cl^-, and OH^-, we write the charge balance of the contents of the titration vessel as

$$[H^+] + [Na^+] = [Cl^-] + [OH^-] \tag{3.3}$$

Note that this relation must be valid at any point during the titration, so that we need not specify the volume V_a+V_b as it changes during the titration.

We now substitute the first two relations into the third. After some rearrangement we obtain

$$\frac{V_b}{V_a} = \frac{C_a - \Delta}{C_b + \Delta} \tag{3.4}$$

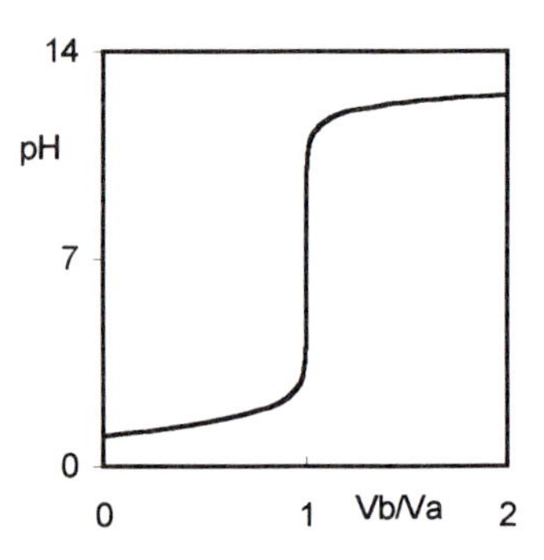

Fig. 3.1b The titration curve for the titration of 0.1 M HCl with 0.1 M NaOH as computed from Eqn. (3.4) with $C_a = C_b = 0.1$. The curve is identical to that of Fig. 3.1a except that the axes have been interchanged.

which is in a form we will use throughout this Chapter. Note that C_a and V_a are the concentration and volume of the *original* sample, while C_b denotes the concentration of the base *in the buret*. All three are therefore constant during the titration, and the only parameter that changes is the volume V_b of base added to the titration vessel. As before, we use Δ as a convenient shorthand notation for $[H^+] - [OH^-] = [H^+] - K_w/[H^+]$.

Equation (3.4) describes the volume V_b added during the titration in terms $[H^+]$ and the constants V_a, C_b, and V_b specifying the particular experimental conditions used. The experimental procedure usually works the other way around: we add a given volume V_b, and record the resulting pH. While, in this particular example, it is easy to invert eqn(3.4) to find an explicit expression for $[H^+]$ in terms of C_a, V_a, C_b, and V_b, in general this approach leads to quite complicated expressions. We will not pursue these here, because our purpose is to explain the general formalism rather than a particular solution of limited applicability.

In chemical experiments we quite often do not interpret our data in precisely the same format in which the experiment is performed, but instead analyze them in a framework that is most convenient to us. For example, we seldom attempt to interpret the measured transmittance of a spectrometric analysis, but rather the resulting absorbance. In that case we let the instrument perform the mathematical *transformation* from transmittance to absorbance for us. Likewise few will use the interferogram of Fourier-transform infrared experiment, or the free induction decay of a Fourier transform nuclear magnetic resonance experiment. In these instances, virtually all users will let the instrument provide the transform which, in this case, is so important that it has lent its name to the entire method.

In the case of acid–base titrations, the transform is about as simple as it can be, because it merely involves an *interchange of the axes*. While such an interchange may seem trivial, we will see below that it allows us to cast the description of titrations in a simple, unifying, general formalism.

Equation (3.4) describes the progress of the titration; we will call the resulting relationship the *progress curve*. We can of course use eqn (3.4) to plot the *titration curve* of pH or $[H^+]$ as a function of V_b, since it makes no difference to an x,y graph whether we compute y as a function of x or the other way around. Progress curves have been used as early as 1940, in Gunnar Hägg's pioneering textbook on the theoretical foundations of analytical chemistry. Figure 3.1 illustrates the progress and titration curves for the titration of HCl with NaOH.

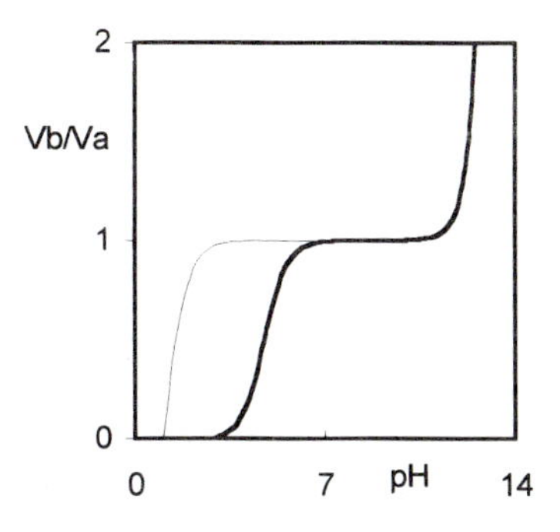

Fig. 3.2a The progress curve for the titration of 0.1 M HAc (pK_a = 4.76) with 0.1 M NaOH as computed from Eqn. (3.8) with C_a = C_b = 0.1. The thin line shows, for comparison, the corresponding curve for HCl from Fig. 3.1a.

We now consider the titration of acetic acid, HAc, a prototypical weak monoprotic acid, with the strong base NaOH. Instead of eqn (3.1) we now have

$$[\mathrm{HAc}]+[\mathrm{Ac}^-]=\frac{C_aV_a}{V_a+V_b} \tag{3.5}$$

but we will need an expression for $[Ac^-]$ rather than for $[HAc] + [Ac^-]$ in the charge balance. We therefore combine eqns (1.20) and (3.5) to

$$[\mathrm{Ac}^-]=\alpha_0\left([\mathrm{HAc}]+[\mathrm{Ac}^-]\right)=\frac{\alpha_0C_aV_a}{V_a+V_b} \tag{3.6}$$

whereupon substitution of eqns (3.2) and (3.6) into the charge balance

$$[\mathrm{H}^+]+[\mathrm{Na}^+]=[\mathrm{Ac}^-]+[\mathrm{OH}^-] \tag{3.7}$$

yields

$$\frac{V_b}{V_a}=\frac{\alpha_0C_a-\Delta}{C_b+\Delta} \tag{3.8}$$

Notice that this expression differs from eqn (3.4) merely in the presence of the multiplier α_0 of C_a, in all other respects leaving the expression unchanged. We can understand this simple result by realizing that the only difference between the right-hand sides of eqns (3.1) and (3.6) lies in the factor α_0. For a strong and therefore fully dissociated acid, $\alpha_0 = 1$, which explains why there is no alpha in eqn (3.4). Fig. 3.2 shows the progress and titration curves for the titration of acetic acid with sodium hydroxide.

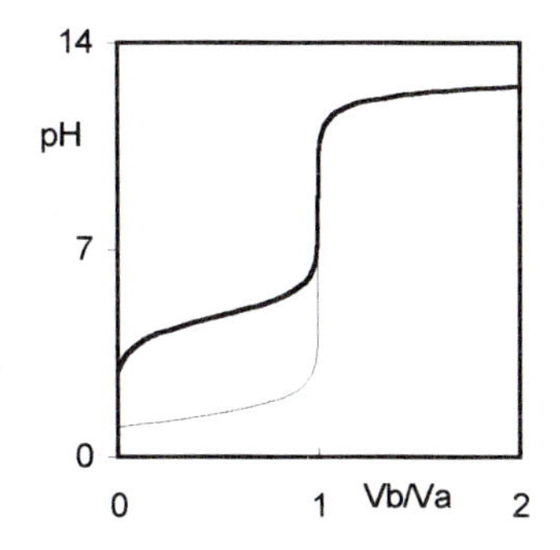

Fig. 3.2b The titration curve for the titration of 0.1 M HAc (pK_a = 4.76) with 0.1 M NaOH as computed from Eqn. (3.8) with C_a = C_b = 0.1. The curve is again identical to that of Fig. 3.2a except that the axes have been interchanged. The thin line shows, for comparison, the corresponding curve for HCl from Fig. 3.1b.

For the titration of a diprotic acid such as oxalic acid, H_2Ox, the mass balance is

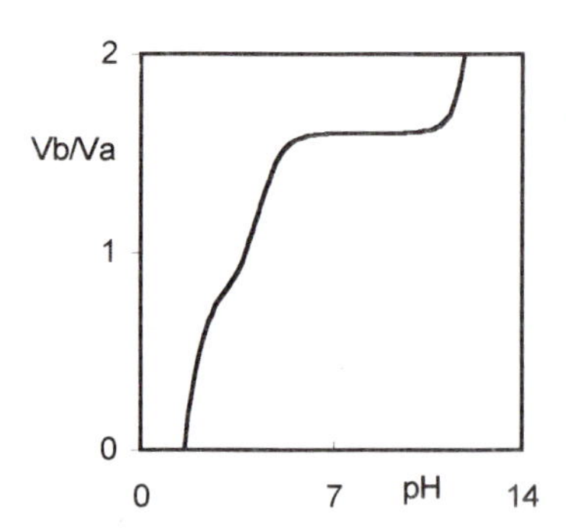

Fig. 3.3a The progress curve for the titration of 0.04 M H_2Ox (pK_{a1} = 1.25, pK_{a2} = 4.27) with 0.05 M NaOH as computed from Eqn. (3.13) with C_a 0.02 and C_b = 0.05.

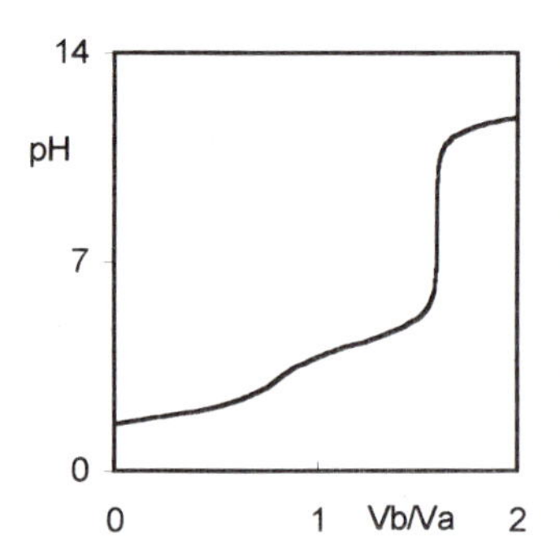

Fig. 3.3b The titration curve for the titration of 0.04 M H_2Ox (pK_{a1} = 1.25, pK_{a2} = 4.27) with 0.05 M NaOH as computed from Eqn. (3.13).

$$[H_2Ox]+[HOx^-]+[Ox^{2-}]=\frac{C_aV_a}{V_a+V_b} \tag{3.9}$$

and the charge balance reads

$$[H^+]+[Na^+]=[HOx^-]+2[Ox^{2-}]+[OH^-] \tag{3.10}$$

where the coefficient 2 precedes the concentration of the divalent oxalate ion, as it should in such an the accounting of charges. We now use

$$[HOx^-]=\alpha_1\left([H_2Ox]+[HOx^-]+[Ox^{2-}]\right)=\frac{\alpha_1C_aV_a}{V_a+V_b} \tag{3.11}$$

and

$$[Ox^{2-}]=\alpha_0\left([H_2Ox]+[HOx^-]+[Ox^{2-}]\right)=\frac{\alpha_0C_aV_a}{V_a+V_b} \tag{3.12}$$

so that the expression for the progress curve now becomes

$$\frac{V_b}{V_a}=\frac{(\alpha_1+2\alpha_0)C_a-\Delta}{C_b+\Delta} \tag{3.13}$$

where we again obtain a result of the same form , but now containing two alphas, one for the titration of H_2Ox to HOx^-, the other for the further titration from HOx^- to Ox^{2-}. Figure 3.3 shows the progress and titration curves.

While eqn (3.13) has a very simple *form*, it reflects a rather complicated dependence of V_b on $[H^+]$, because both α_1 and α_0 are functions of $[H^+]$, see eqns (1.25) and (1.26), and so is Δ. But this cannot be helped: the simplicity is in the formalism, the complexity in the chemistry (the two successive deprotonation steps of the acid, plus the autodissociation of water), and this is how it should be.

For the titration of a triprotic acid such as H_3PO_4, with mass balance

$$[H_3PO_4]+[H_2PO_4^-]+[HPO_4^{2-}]+[PO_4^{3-}]=\frac{C_aV_a}{V_a+V_b} \tag{3.14}$$

and charge balance

$$[H^+]+[Na^+]=[H_2PO_4^-]+2[HPO_4^{2-}]+3[PO_4^{3-}]+[OH^-] \tag{3.15}$$

we likewise find

$$\frac{V_b}{V_a}=\frac{(\alpha_2+2\alpha_1+3\alpha_0)C_a-\Delta}{C_b+\Delta} \tag{3.16}$$

with three alphas, one each for the three stages of this titration, from H_3PO_4 through HPO_4^- and HPO_4^{2-} to PO_4^{3-}. Figure 3.4 illustrates this result.

Comparison of the above results with the formalism introduced earlier suggests the following generalizations:

1. the factor multiplying C_a is the acid dissociation function F_a, and
2. the right-hand side of eqns (3.4), (3.8), (3.13), and (3.16) is always the ratio of the proton conditions of the sample and titrant respectively, each written in its standard form.

We will not prove these inferences here, but they are indeed correct, and provide a general framework for acid–base titrations.

We can therefore write the general expression for the progress of the titration of *any* acid with a single strong monoprotic base as

$$\frac{V_b}{V_a} = \frac{F_a C_a - \Delta}{C_b + \Delta} \tag{3.17}$$

We had seen earlier that the proton condition of a mixture can be written as the algebraic sum of the contributions of the mixture components plus that of water (the term Δ), and the same applies to the titration of such a mixture. For example, for the titration of an arbitrary mixture of acids we obtain

$$\frac{V_b}{V_a} = \frac{\sum_i F_{ai} C_{ai} - \Delta}{C_b + \Delta} \tag{3.18}$$

and the titration of any mixture of i acids with any mixture of j bases by

$$\frac{V_b}{V_a} = \frac{\sum_i F_{ai} C_{ai} - \Delta}{\sum_j F_{bj} C_{bj} + \Delta} \tag{3.19}$$

while the reverse is also true: the titration of a mixture of any number j of bases with a mixture of any number i of acids is described completely by

$$\frac{V_a}{V_b} = \frac{\sum_j F_{bj} C_{bj} + \Delta}{\sum_i F_{ai} C_{ai} - \Delta} \tag{3.20}$$

which you will recognize as the inverted form of eqn (3.19).

The titration of a mixture of acids or bases is not simply the sum of the titrations of the individual components of such a mixture. Equations (3.19) and (3.20) show at what level these individual processes are additive in the formalism.

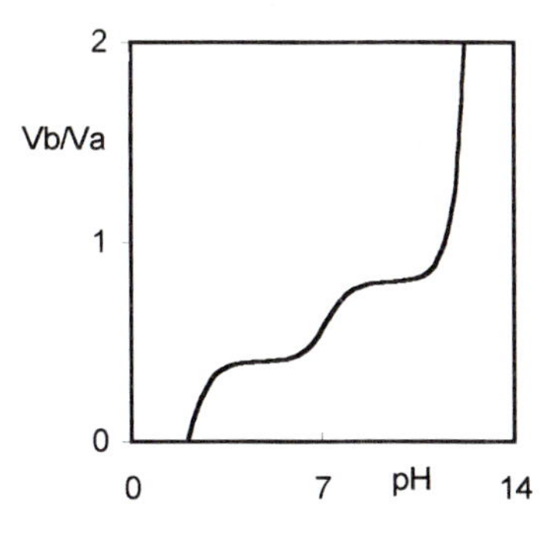

Fig. 3.4a The progress curve for the titration of 0.02 M H_3PO_4 (pK_{a1} = 2.15, pK_{a2} = 7.20, pK_{a3} = 12.15) with 0.05 M NaOH as computed from eqn (3.16). The first equivalence point is at $C_aV_a = C_bV_b$ or $V_b/V_a = C_a/C_b = 0.4$, and the 2nd and 3rd equivalence points are at V_b/V_a = 0.8 and 1.2 respectively.

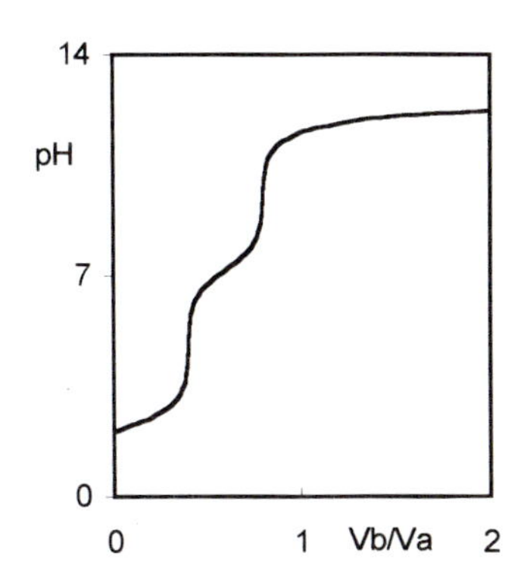

Fig. 3.4b The titration curve for the titration of 0.02 M H_3PO_4 (pK_{a1} = 2.15, pK_{a2} = 7.20, pK_{a3} = 12.15) with 0.05 M NaOH as computed from eqn (3.16).

3.2 Examples

Here we will illustrate the application of the relations derived so far by considering the titration of sulfuric acid with a strong base such as NaOH, and the titration of a mixture of carbonate and bicarbonate with a strong acid such as HCl. Sulfuric acid acts as a strong diprotic acid in dilute solutions, but as a (not quite resolved) combination of a strong and a weak monoprotic acid at high concentrations. The carbonate/bicarbonate titration illustrates how to deal with bases as well as with mixtures.

3.2a The titration of H_2SO_4 with NaOH

Sulfuric acid has two protons; the first is completely dissociated in water, whereas the second deprotonation is characterized by a pK_a of about 2. Formally, the progress curve for the titration of C_a M H_2SO_4 with C_b M NaOH is given by eqn (3.13), but this expression now contains the unknown value of K_{a1}. One could use the equation by substituting a sufficiently large value for K_{a1}, but a more elegant way is to divide through K_{a1} and then to let K_{a1} go to infinity,

$$\alpha_1 + 2\alpha_0 = \frac{[\mathrm{H}^+]K_{a1} + 2K_{a1}K_{a2}}{[\mathrm{H}^+]^2 + [\mathrm{H}^+]K_{a1} + K_{a1}K_{a2}}$$

$$= \frac{[\mathrm{H}^+] + 2K_{a2}}{[\mathrm{H}^+]^2 / K_{a1} + [\mathrm{H}^+] + K_{a2}} \approx \frac{[\mathrm{H}^+] + 2K_{a2}}{[\mathrm{H}^+] + K_{a2}} = 1 + \frac{K_{a2}}{[\mathrm{H}^+] + K_{a2}} \quad (3.21)$$

so that the titration proceeds as if it were an equimolar mixture of a strong and weak acid. Figure 3.5 illustrates the resulting titration curves for concentrated and dilute samples.

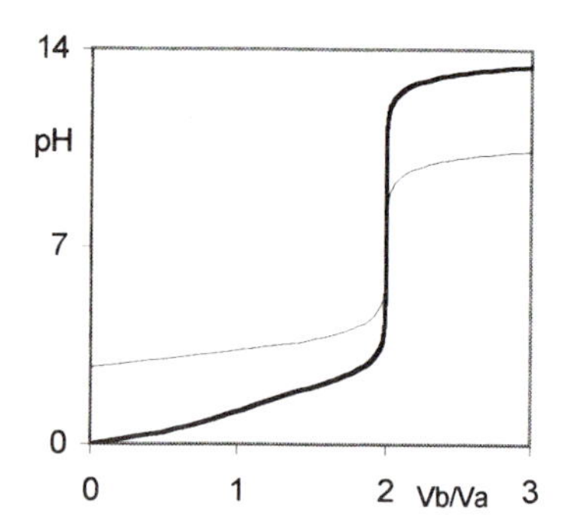

Fig. 3.5 The titration curve for the titration of 1 M H_2SO_4 with 1 M NaOH (heavy line) or that of 1 mM H_2SO_4 with 1 mM NaOH (thin line).

3.2b The simultaneous titration of Na_2CO_3 and $NaHCO_3$ with HCl

The progress curve for the titration with a strong monoprotic acid of a diprotic base such as sodium carbonate is defined by

$$\frac{V_a}{V_b} = \frac{(\alpha_1 + 2\alpha_2)C_b + \Delta}{C_a - \Delta} \quad (3.22)$$

which is the counterpart of eqn (3.13), and the corresponding expression for bicarbonate is

$$\frac{V_a}{V_b} = \frac{(\alpha_2 - \alpha_0)C_b + \Delta}{C_a - \Delta} \quad (3.23)$$

so that the titration with HCl of a mixture containing C_{b1} M Na_2CO_3 + C_{b2} M $NaHCO_3$ will be given by the expression

$$\frac{V_a}{V_b} = \frac{(\alpha_1 + 2\alpha_2)C_{b1} + (\alpha_2 - \alpha_0)C_{b2} + \Delta}{C_a - \Delta}$$

$$= \frac{\dfrac{\left(2[\mathrm{H}^+]^2 + [\mathrm{H}^+]K_{a1}\right)C_{b1} + \left([\mathrm{H}^+]^2 - K_{a1}K_{a2}\right)C_{b2}}{[\mathrm{H}^+]^2 + [\mathrm{H}^+]K_{a1} + K_{a1}K_{a2}} + \Delta}{C_a - \Delta} \quad (3.24)$$

as illustrated for various combinations of C_{b1} and C_{b2} in Fig. 3.6.

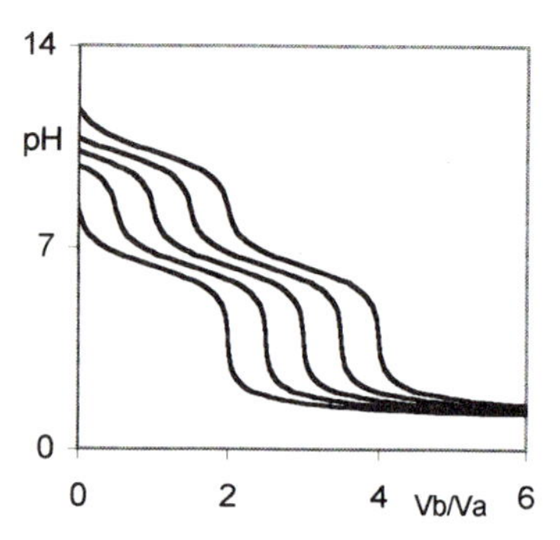

Fig. 3.6 Titration curves for the titration of solutions containing C_{b1} M Na_2CO_3 + C_{b2} M $NaHCO_3$ with 0.1 M HCl. From top to bottom: C_{b1} = 0.20, 0.15, 0.10, 0.05, and 0.00 M; C_{b1} = 0.00, 0.05, 0.10, 0.15, and 0.20 M. Consequently the top and bottom curves are for 0.2 M Na_2CO_3 and 0.2 M $NaHCO_3$ only.

3.3 Determining the equivalence point

The primary purpose of an analytical titration is to determine the equivalence point or, more precisely, the corresponding titrant volume V_e. For the titration of a single, concentrated strong acid and base, such as depicted in Fig. 3.1, the location of the equivalence point is obvious, and is characterized by a very large change d(pH)/dV_t, where V_t is the volume of titrant added. In fact, methods to determine the equivalence point are sometimes based on this derivative, d(pH)/dV_t, see below in paragraph 3.2b.

In other cases, the equivalence point is more difficult to pin down. Fortunately there are a number of different methods to determine the equivalence point. We will here describe several of these, and compare them critically.

3.3a Methods for detecting a specific pH

Since titrations are primarily used as quantitative tools, one will usually have a fair estimate of the pH at the equivalence point. In that case one can simply use a pH meter equipped with an alarm system to alert the user that a particular pH value has been crossed. This can easily be automated, so that the titration stops at that point, and starts a sequence in which the dispensed titrant volume is recorded, the titration vessel cleaned, a new sample introduced, and a new titration started. There is nothing wrong with this method, as long as it operates in a region where the titration curve is so steep that errors due to noise in the pH readings (typically mostly from insufficiently fast mixing of the added titrant) and drift in the pH meter remain insignificant.

An equivalent method is to use a color indicator, although in this case the endpoint indication is much less sharp (since the indicator color changes gradually) and the options for setting a particular equivalence point pH are limited by the available indicator pK_a-values. Moreover, the indicator is itself titrated, and therefore should be present only at a concentration much smaller than that of the sample, a factor that restricts indicators to highly colored species, i.e., to those with molar absorptivities of the order of at least 10^4 M^{-1} cm^{-1}. For the above reasons, the use of visual color indicators is usually inferior in precision to using a pH-meter as a level-crossing detector. On the other hand, color indicators have the advantage that they require no instrumentation.

3.3b Methods for detecting another equivalence point property

The pH at the equivalence point is a (weak) function of the sample concentration, and therefore cannot be known precisely in advance of the titration. A method that does not have this problem, and is also not influenced by drift in the pH meter, is to use the first or second derivative of the titration curve, $d(pH)/dV_t$ or $d^2(pH)/dV_t^2$. In that case it is customary to associate the equivalence point with the maximum value in $d(pH)/dV_t$ or the zero-crossing of $d^2(pH)/dV_t^2$. Meites and Goldman showed that the maximum in $d(pH)/dV_t$ does *not* coincide exactly with the equivalence point, although the difference is often negligibly small. A more serious disadvantage of these methods is that they are very sensitive to noise. Unfortunately, the dominant noise in an acid–base titration is due to incomplete mixing, and is maximal at the equivalence point, where the buffer strength of the test solution is minimal. The process of taking the derivative further enhances the noise. The method is therefore not very robust.

3.3c Methods using a subset of the titration curve

The most prominent methods in this category are those of Gran, and use linear extrapolation of the data before or after the equivalence point to find the equivalence volume V_e. They are based on approximations to the titration curves of monoprotic acids and bases either *before* of *after* the equivalence point, as listed in Table 3.1. Below we will illustrate how such relations can be derived from the general titration equations.

For that part of the titration of a strong acid with a strong base that lies before the equivalence point we start from eqn (3.4) with the approximation $\Delta = [H^+] - [OH^-] \approx [H^+]$ and obtain

$$\frac{V_b}{V_a} = \frac{C_a - [H^+]}{C_b + [H^+]} \tag{3.25}$$

which can be rearranged to

$$[H^+]\,(V_a + V_b) = C_a V_a - C_b V_b$$

$$= (C_a V_a / C_b - V_b)\,C_b = (V_e - V_b)\,C_b \tag{3.26}$$

and then forms the basis of the Gran1 plot.

For the part past the equivalence point, we combine eqn (3.4) with the approximation $\Delta = [H^+] - [OH^-] \approx -[OH^-]$ to

$$\frac{V_b}{V_a} = \frac{C_a + [OH^-]}{C_b - [OH^-]} \tag{3.27}$$

which leads to

$$[OH^-]\,(V_a + V_b) = C_b V_b - C_a V_a$$

$$= (C_b V_b - C_a V_a) = (V_b - V_e)\,C_b \tag{3.28}$$

or, after substitution of $[OH^-] = K_w/[H^+]$,

$$(V_a + V_b)/[H^+] = (V_b - V_e)\,C_b / K_w \tag{3.29}$$

which underlies the Gran2 plot. Both of these Gran plots for the titration of a strong acid with a strong base, or the corresponding expressions for the titration of a strong base with a strong acid, are useful as long as the concentrations of acid and base are not too low.

For the titration of a weak monoprotic acid with a strong monoprotic base we start with eqn (3.8) which, after substitution of eqn (1.23), reads

$$\frac{V_b}{V_a} = \frac{C_a K_a / ([H^+] + K_a) - \Delta}{C_b + \Delta} \tag{3.30}$$

In order to reduce eqn (3.30), which is third order in $[H^+]$, to an expression that is linear in $[H^+]$, one now makes the more drastic assumptions $\Delta \ll \alpha_0 C_a$ and $\Delta \ll C_b$ so that eqn (3.30) reduces to

$$\frac{V_b}{V_a} \approx \frac{C_a K_a}{C_b\,([H^+] + K_a)} \tag{3.31}$$

which after some algebraic manipulation yields

$$[H^+]\,V_b = (C_a V_a / C_b - V_b)\,K_a = (V_e - V_b)\,K_a \tag{3.32}$$

which forms the basis of the Gran1 plot for the titration a weak acid with a strong base before the equivalence point. We note that the approximation $\Delta \ll \alpha_0 C_a$ made in deriving eqn (3.32) makes it unsuitable for use with fairly strong acids. Indeed, eqn (3.32) does not approach eqn (3.26) when we let K_a increase. Beyond the equivalence point, one can use eqn (3.28) or (3.29).

Table 3.1 presents the relevant expressions, while Figs. 3.7 show how well these Gran plots apply. In general, they work well for fairly concentrated strong acids and bases, and moderately well for sufficiently concentrated weak acids and bases.

Table 3.1 The Gran relations for the titration of a single (weak or strong) monoprotic acid with a strong base, or vice versa.

Titration with a strong monoprotic base of a	*Gran 1: before the equivalence point*	*Gran 2: beyond the equivalence point*
weak monoprotic acid:	$[H^+]\,V_b = (V_e - V_b)\,K_a$	$(V_a + V_b)\,/\,[H^+] = (V_b - V_e)\,C_b\,/\,K_w$
strong monoprotic acid:	$[H^+]\,(V_a + V_b) = (V_e - V_b)\,C_b$	
Titration with a strong monoprotic acid of a	*Gran 1: before the equivalence point*	*Gran 2: beyond the equivalence point*
weak monoprotic base:	$V_a\,/\,[H^+] = (V_e - V_a)\,/\,K_a$	$[H^+]\,(V_a + V_b) = (V_a - V_e)\,C_a$
strong monoprotic base:	$(V_a + V_b)\,/\,[H^+] = (V_e - V_a)\,C_a\,/\,K_w$	

With polyprotic acids and bases, Gran plots are often insufficiently linear, and sometimes fail to extrapolate to the equivalence point, except for the second Gran plot (past the last equivalence point) which usually remains robust because it is based on the acid–base properties of the titrant, typically either a strong acid or a strong base. In the titration of weak acids or bases, extrapolation of Gran1 plots is often necessary because they can be noticeably curved near the equivalence point.

3.3d Methods that use the entire titration curve

Schwartz has shown that, for the titration of a single (strong or weak) monoprotic acid or base with its strong counterpart, it is possible to linearize the entire titration curve *without* making simplifying assumptions. Starting from eqn (3.30) we write

$$C_b V_b + V_b \Delta = C_a V_a K_a / ([\mathrm{H}^+] + K_a) - V_a \Delta \tag{3.33}$$

$$V_b + (V_a + V_b)\Delta / C_b = \frac{C_a V_a}{C_b} \frac{K_a}{[\mathrm{H}^+] + K_a} = \frac{V_e K_a}{[\mathrm{H}^+] + K_a} \tag{3.34}$$

which leads directly to the expressions for this case in Table 3.2.

Figure 3.7a compares the various methods for a theoretical titration curve, and Fig. 3.7b does the same for a titration curve containing some added Gaussian noise. We see that the Schwartz plot is somewhat less robust

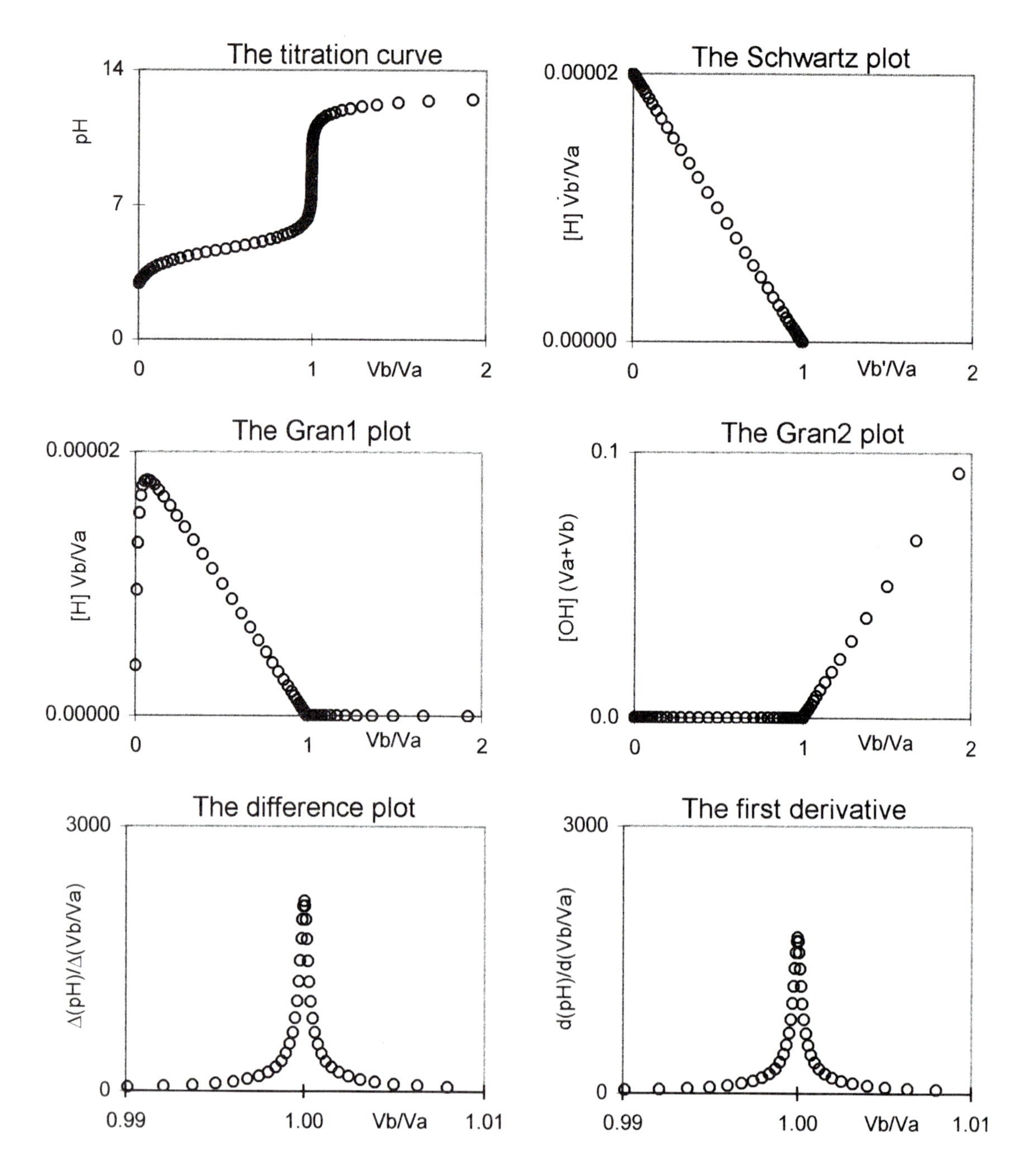

Fig. 3.7a. The titration curve (top left) calculated for the titration of 0.1 M acetic acid (pK_a = 4.76) with 0.1 M NaOH. Also shown are the corresponding Schwartz plot (top right), the Gran1 and Gran2 plots (left and right center), the difference plot $\Delta(pH)/\Delta(V_b/V_a)$ (bottom left) and the first derivative plot $d(pH)/d(V_b/V_a)$; the latter is calculated using a Savitzky–Golay thirteen-point quadratic.

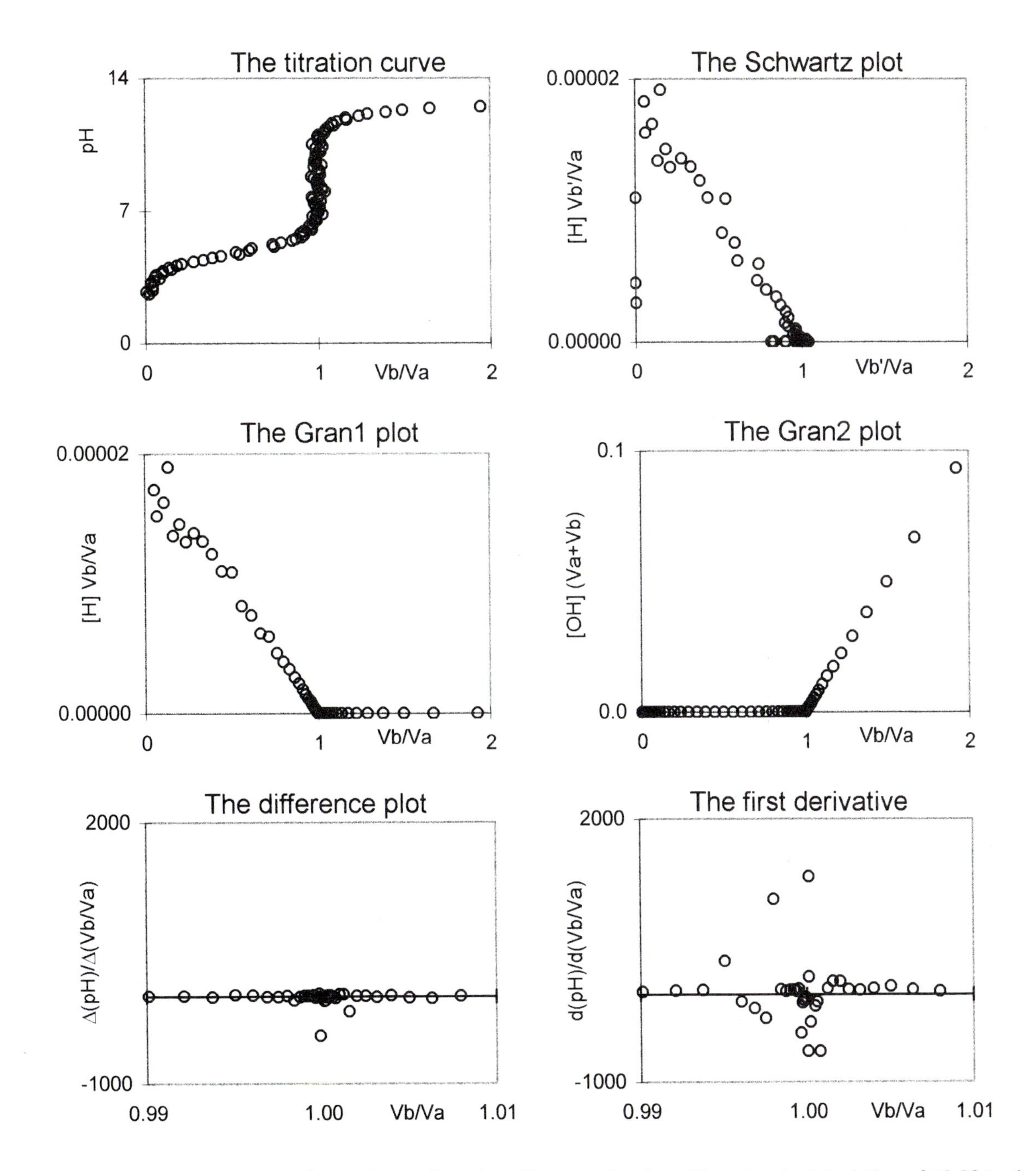

Fig. 3.7b. The same curves after addition of random (Gaussian) noise with a standard deviation of ±0.02 to the calculated values of V_b/V_a. The Gran2 (and, to a lesser extent, the Gran1) plots are quite tolerant of such noise, while the derivative and difference plots clearly are not.

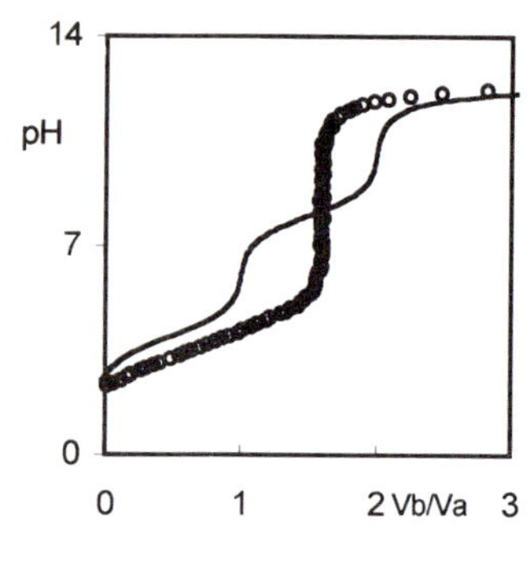

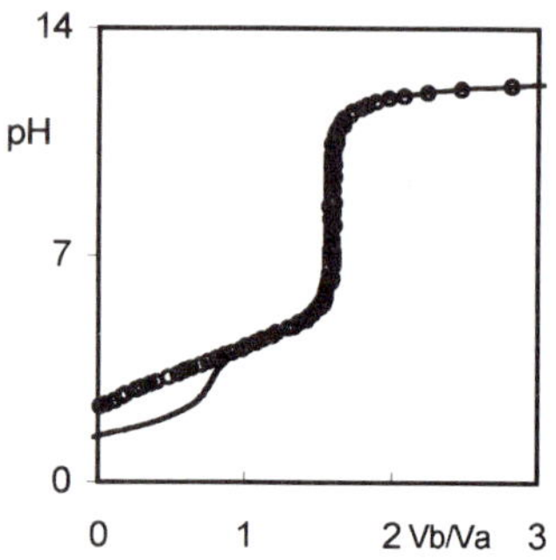

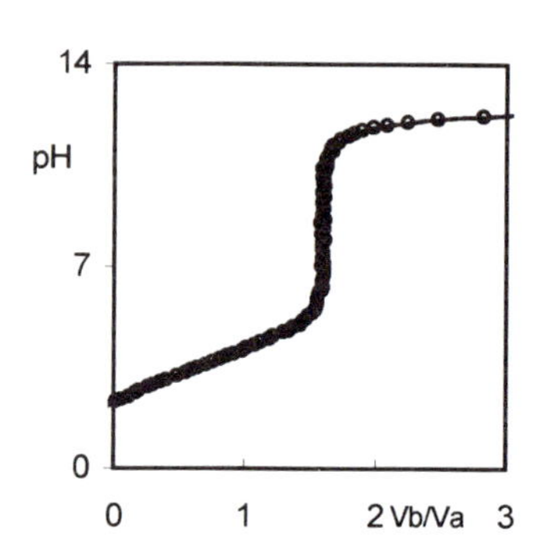

Fig. 3.8 Three successive stages in the non-linear least-squares fitting of the titration curve for the titration of 0.04 M fumaric acid (pK_{a1} = 3.05, pK_{a2} = 4.49) with 0.05 M NaOH, as computed from eqn (3.16) with Gaussian noise (with a standard deviation of ±0.01) added to both axes (open circles). Starting assumptions: C_a = 0.05 M, pK_{a1} = 4.00, pK_{a2} = 8.00; final values C_a = 0.040 M, pK_{a1} = 3.07, pK_{a2} = 4.51. Less noise will yield a better fit.

than the first Gran plot, which in turn is less robust than the second Gran plot. While it is possible to design properly weighted least squares algorithms to fit Gran and Schwartz plots while discriminating against the noisy parts of these plots, these methods are still inapplicable to polyprotic acids and bases with closely-spaced pK_a-values, and to mixtures.

Table 3.2 The Schwartz relations for the titration of a single (weak or strong) monoprotic acid with a strong base, or vice versa.

titration of an acid with a strong base:	$[H^+]\,V_b' = (V_e - V_b')\,K_a$ $V_b' = V_b + (V_a + V_b)\,\Delta\,/\,C_b$ $\Delta = [H^+] - [OH^-]$
titration of a base with a strong acid:	$V_a'\,/\,[H^+] = (V_e - V_a')\,/\,K_a$ $V_a' = V_a - (V_a + V_b)\,\Delta\,/\,C_a$

The most robust and by far the most general method available so far appears to be a non-linear least-squares fit to the entire titration curve. This method only requires that one knows what *type* of sample to expect: one clearly cannot expect to fit the titration of a triprotic acid with the equation for a diprotic one.

However, this method has some peculiar pitfalls, related to the properties of the theoretical expressions at physically nonrealizable pH regions. When these problems are avoided (as they always can), this appears to be a very powerful general method of fitting titration data.

Here is the problem: the expressions for V_b/V_a or V_a/V_b developed in Section 3.1 have no inherent constraints on the values of $[H^+]$. However, any real, physical system does have such constraints: 0.1 M acetic acid has a pH of 2.88, and its titration with 0.1 M NaOH can never bring the pH below 2.88 or above 13, the pH of the undiluted titrant. The expression for V_b/V_a will yield *negative* values when used with pH-values smaller than 2.88 or larger than 13. While these would alert the user that the computed results are not physically realizable, such non-realizable values can be taken at face value by a naïve least-squares routine. Moreover, since the deviant values can be quite large, they may dominate the least-squares fit, which is extra sensitive to large deviations, since it tries to minimize their even larger *squares*. These problems can be avoided by judicious data analysis, as described in detail in my book *Excel in Analytical Chemistry* (de Levie 1999c). When that is done, a general data fitting routine results which is remarkably insensitive to noise, and applicable to any acid–base titration problem. Figure 3.8 illustrates the noise-tolerance of this method.

As already indicated, there are uses of titrations beyond concentration determinations. The most prominent of these is the determination of equilibrium constants. Such constants cannot be obtained from the equivalence points, but require analysis of data in the so-called buffer regions where $d(pH)/dV_t$ is small. Gran and Schwartz plots can be used to generate values for K_a, insofar as one deals with either monoprotic acids or bases, or with polyprotic acids and bases with widely spaced pK_a-values, so that the various neutralization processes are essentially uncoupled from each other, and therefore appear as well-separated processes in the titration curve. The most general method is, again, a non-linear least-squares fit of the data to an appropriate expression for the progress of the titration, in which case the computer program will determine the best-fitting values of the equilibrium constants. As already mentioned in Section 2.7, many special programs have also been written for this purpose.

3.4 Diagrams and titrations

It is useful to try to correlate the progress or titrations curves to the logarithmic concentration diagram of the sample. Strictly speaking this cannot be done, since the logarithmic concentration diagram reflects the properties of a single solution, whereas a titration necessarily involves two different solutions that are being mixed and then react during the titration. In practice, however, we will typically select as titrant a single strong acid or base. Moreover, dilution effects are usually small, and certainly can be made to be so by using a titrant that is much more concentrated than the sample. Under such simplifying conditions, we can look at the logarithmic concentration diagram of the sample to see whether a titration is possible. If a titration is not feasible under those favorable conditions, it certainly will not be possible when dilution is taken into account, when the titrant is a weak acid or base, etc.

Consider the logarithmic concentration diagram of 0.1 M acetic acid in Figs. 1.7, 2.1, and 2.2. A solution of 0.1 M acetic acid has a pH of 2.88, as indicated by point *b* in Fig. 2.1. Halfway to the equivalence point, about half of the acetic acid has been deprotonated, i.e., $[\mathrm{HAc}] \approx [\mathrm{Ac}^-]$ so that pH $\approx$ pK_a = 4.76. At the equivalence point all acetic acid has been reacted to sodium acetate, and (apart from dilution) the pH will therefore be 8.88, see point *d* in Fig. 2.2. At 100% excess of titrant (and again neglecting dilution effects) we have a solution 0.1 M in NaAc and 0.1 M in NaOH, so that the pH will be 13, point *e* in Fig. 2.2. Therefore we see that the entire pH swing near the equivalence point must be contained between points *c* (at 50% completion of the titration) and *e* (at 200% completion) in Fig. 2.2. We can use that distance as a rough measure of the *steepness* of the titration. In fact, it is more convenient to take a related distance that is uniquely related to the equivalence point, and it is customary to use as such the vertical distance between the line at $\log c = \log C$ and the particular point in the diagram that represents the equivalence point, i.e., point *d* in Fig. 2.2. That vertical distance is precisely half that between points *c* and *e*. We will therefore define the *titration steepness* S_t as $C/[\mathrm{H}^+]_e$, where $[\mathrm{H}^+]_e$ is the value of $[\mathrm{H}^+]$ at the equivalence point.

Figure 3.9 shows several combinations of logarithmic concentration diagrams and titration curves to illustrate that S_t as just defined indeed correlates

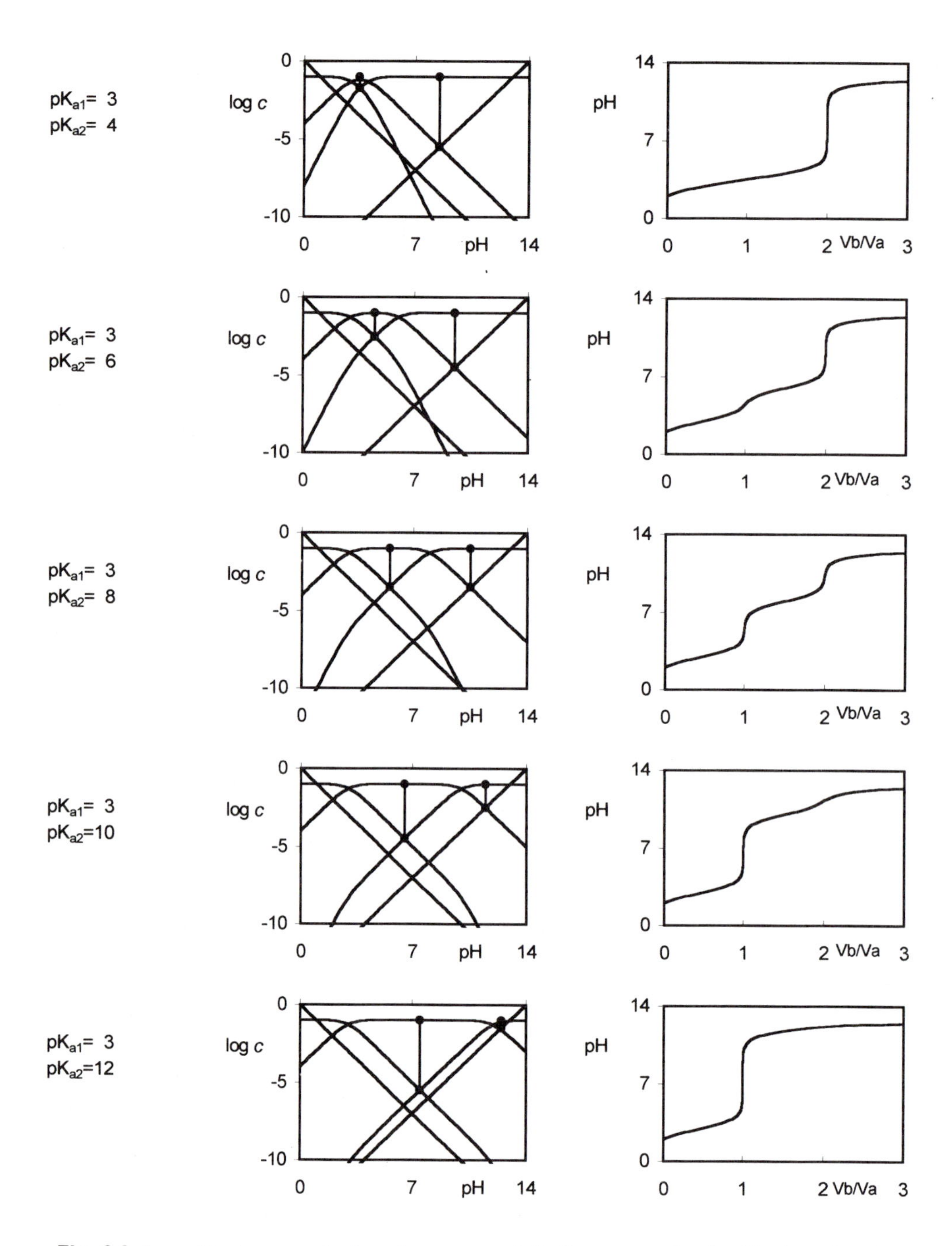

Fig. 3.9. Logarithmic concentration diagrams and titration curves for the titration of a single diprotic weak acid with a single strong monoprotic base, both at 0.1 M. The pK_a-values of the diprotic acid are listed to the left of the plots. The vertical lines between heavy dots show the steepness S_t of the titration curve.

with the steepness of the titration. In general, $S_t \geq 10^3$ indicates the possibility of an accurate titration, a value of S_t between 10^2 and 10^3 corresponds to a feasible titration of limited accuracy (of between 1% and 0.1%), while titrations with S_t below 10^2 should be avoided.

Figures 3.10 and 3.11 illustrate how the steepness as estimated from a logarithmic concentration diagram can guide us in determining what equivalence point(s) to use for a polyprotic acid or base. In Figures 2.21 and 3.10 we see that the titration of phosphoric acid to its third equivalence point has virtually no steepness; indeed, one usually cannot even observe any change of slope in the titration curve at that point, see Fig. 3.4b. Titrating to the first equivalence point (s = 1 when the titrant is a monoprotic base) is fine, and titrating to the second equivalence point (s = 2) is equally satisfactory, but will consume more titrant. With citric acid the situation is just the reverse: the first two equivalence points are useless, as one can anticipate from the logarithmic concentration diagram in Fig. 3.11a. The steepness at the third equivalence point is fine, and titrating to that third equivalence point is indeed very suitable, see Fig. 3.11b. If the titrant in that case is NaOH, the value of the stoichometric factor s is 3; if the titrant were, say, $Mg(OH)_2$, we would have $s = 3/2$.

We can also use logarithmic concentration diagrams to explore how to titrate acid salts, which in principle can be titrated with either acid or base. Figure 3.12 shows the logarithmic concentration diagram for phthalic acid. Titrating potassium hydrogen phthalate with strong acid would get us to point a in the diagram, with little steepness; titrating with strong base yields equivalence point b with a titration steepness S_t of the order of 10^4, sufficient for use as a primary standard for the standardization of strong bases.

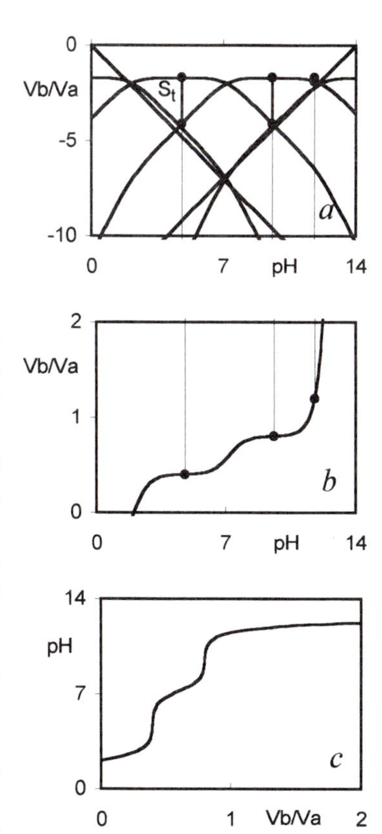

Fig. 3.10 The logarithmic concentration diagram (a) of 0.02 M H_3PO_4 (pK_{a1} = 2.15, pK_{a3} = 7.20, pK_{a3} = 12.15), and the corresponding progress curve (*b*) and titration curve (*c*). The steepness S_t of the titration curve is indicated by heavy lines in the top graph, while thin lines indicate the locations of the equivalence points in the top two panels, which share a common pH axis. Only the first and second equivalence points (s = 1 or 2) are useful, as can be seen from the steepness S_t (as indicated in Fig. 3.10a as the vertical distance between two heavy dots) in the logarithmic concentration diagram.

3.5 Titration errors

As in any quantitative method, there are several possible sources of errors that must be avoided in order to realize results of high accuracy and precision. These sources of error depend on the method used.

When we use a color indicator, the dominant errors are so-called *indicator errors*, resulting from a mismatch between the equivalence point pH and the pK_a of the indicator used, and from the width of the pH range over which the color changes. The formalism of Section 3.1 makes it simple to estimate the resulting systematic (determinate) errors, because we use the pH values bracketing the indicator range, and compute the resulting values of V_t/V_s.

For example, consider the titration of 0.1 M acetic acid with 0.1 M NaOH, using as indicator phenolphthalein, with a reported transition range from pH 8.2 to pH 9.8. Substituting $K_a = 10^{-4.76}$, and $[H^+] = 10^{-8.2}$ into eqns (1.23) and (3.8) yields $V_b/V_a = 0.9997$, while we find $V_b/V_a = 1.0013$ for $[H^+] = 10^{-9.8}$, so that the resulting errors are of the order of 0.1% and therefore quite acceptable. However, when we repeat this exercise for $C_a = C_b$ = 1 mM, we calculate $V_b/V_a = 1.0030$ and $V_b/V_a = 1.1347$ respectively, because the pH of the equivalence point has now moved outside the indicator range.

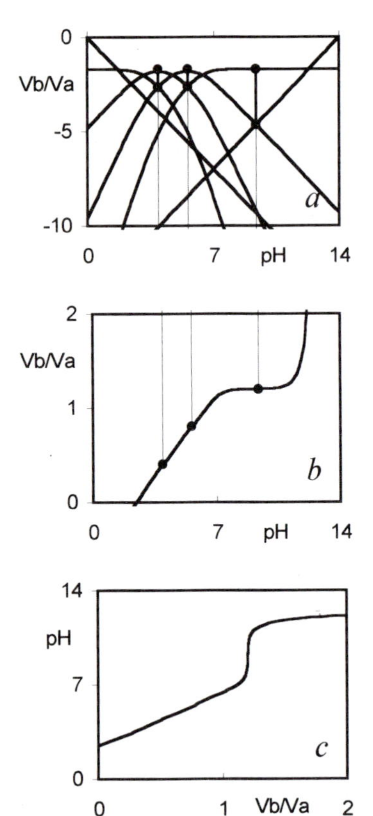

Fig. 3.11 The logarithmic concentration diagram (*a*) of 0.02 M citric acid (pK_{a1} = 3.13, pK_{a3} = 4.76, pK_{a3} = 6.40), and the corresponding progress curve (*b*) and titration curve (*c*) for its titration with 0.05 M NaOH. Comparison with Fig. 3.8 illustrates how the logarithmic concentration diagram shows why we should now titrate to the third equivalence point, *s* = 3.

A possible maximal titration error of 13.5% is clearly unacceptable, and for the latter titration one should use a more appropriate indicator, with a range around the equivalence point pH of about 7.88, such as cresol purple.

With potentiometric titrations, improper design of the titration cell, and specifically the relative placements of the stirrer, buret tip, and sensing pH electrode in that cell, are often a dominant source of both lag and noise in the pH readings, especially in an automated titration. As can be seen in Fig. 3.7b, the various analysis methods differ greatly in their sensitivity to noise, i.e., to random (indeterminate) errors. Moreover, offset due to drift or lack of proper calibration can also lead to significant systematic errors, and likewise affects the various methods to different extents. For example, while methods based on the first or second derivative are most sensitive to noise, they are rather insensitive to the effects of drift, and totally immune to the effects of offset. On the other hand, Gran and Schwartz plots involve computing the proton concentration from the measured pH, a calculation that can introduce rather large errors even for pH readings that are only slightly off. As with all quantitative methods, high-quality results are obtained only with proper experiment design, continuous vigilance, and repeated calibration.

3.6 The traditional approach

The treatment of titrations given so far is quite different from that found in most textbooks. For the sake of completeness we will here summarize the more traditional approach, and then use the opportunity to highlight the differences.

First we reconsider the titration of HCl with NaOH as the prototypical neutralization of a strong acid by a strong base. We subdivide the problem into two parts, i.e., the curve before and after the equivalence point.

For the section of the titration curve between the start of the titration and the equivalence point we calculate the proton concentration as follows. The total amount of acid is C_aV_a, and after addition of a volume V_b of base the remaining amount of excess acid is $C_aV_a - C_bV_b$. Since the volume of the partially titrated sample is now $V_a + V_b$, the hydrogen concentration is

$$[H^+] = \frac{C_aV_a - C_bV_b}{V_a + V_b} \quad \text{before the equivalence point} \qquad (3.35)$$

For the section of the titration curve past the equivalence point, the same logic is applied to the excess amount of base is $C_bV_b - C_aV_a$, so that we obtain

$$[OH^-] = \frac{C_bV_b - C_aV_a}{V_a + V_b} \quad \text{beyond the equivalence point} \qquad (3.36)$$

or

$$[H^+] = \frac{K_w}{[OH^-]} = \frac{K_w(V_a + V_b)}{C_bV_b - C_aV_a} \quad \text{past the equivalence point} \qquad (3.37)$$

At the equivalence point we have $C_aV_a = C_bV_b$, and at that point eqn

(3.35) yields $[H^+] = 0$ while eqn (3.36) leads to $[OH^-] = 0$ and eqn (3.37) to $[H^+] = \infty$. Such results are obviously erroneous (the correct answer being $[H^+] = [OH^-] = \sqrt{K_w} = 10^{-7}$ M), and in this approach the equivalence point is therefore treated separately. We note that eqn (3.35) is identical to eqn (3.26), and that, likewise, eqns (3.36) and (3.37) are fully equivalent to eqns (3.28) and (3.29).

Next we consider the titration of a weak acid with a strong base. For the part of the titration curve before the equivalence point, one often neglects the autodissociation of the weak acid, and furthermore assumes that the amount of acid anions is fully equivalent to the amount of base added during the titration. Such reasoning, which does not show precisely what assumptions are made, but is equivalent to those underlying eqn (3.32), leads to

$$[H^+] = K_a \frac{C_a V_a - C_b V_b}{C_b V_b} \quad \text{before the equivalence point} \qquad (3.38)$$

while the part beyond the equivalence point is assumed to be given by eqn (3.36) or (3.37).

In this case there is an additional problem: not only does eqn (3.38) yield $[H^+] = 0$ at the equivalence point, but it also gives an incorrect answer for the onset of the titration, where $V_b = 0$ so that eqn (3.38) goes to infinity. (The identical Gran expression (3.32) has the same problem, which is why the initial part of the Gran1 plot for the titration of a weak acid must always be excluded from the linear analysis.) So now we need two points, at the beginning of the titration and at the equivalence point, and two approximate segments that neither join nor pass through those points.

Perhaps this is also why misconceptions, such as that the maximum in d(pH)/dV_t coincides with the equivalence point, can survive. Since neither curve section goes through the equivalence point, and methods to determine the derivative of a single, isolated point have yet to be developed, such a (false) statement cannot be falsified!

Quantitative comparison of eqn (3.38) with the exact result shows that eqn (3.38) can be quite far off, especially for not very weak acids, as illustrated in Fig. 3.13.

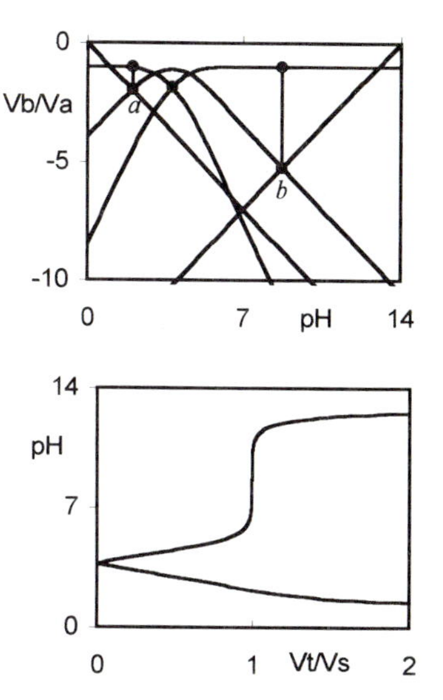

Fig. 3.12 The logarithmic concentration diagram of 0.1 M phthalic acid (pK_{a1} = 2.95, pK_{a1} = 5.41) and the corresponding titration curves of 0.1 M potassium hydrogen phthalate with 0.1 M HCl (bottom curve) and 0.1 M NaOH (top curve) respectively. Again, the logarithmic concentration diagram shows which titration is feasible.

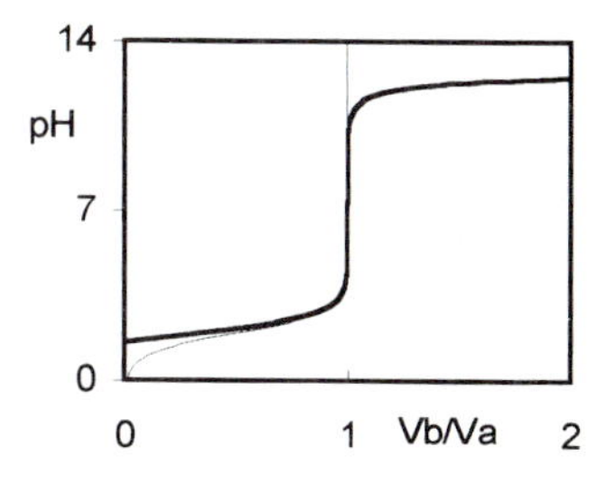

Fig. 3.13 The titration curve for the titration of 0.1 M chlorous acid (pK_a = 1.95) according to eqn (3.8) (heavy line) and its approximation, eqn (3.34) or $V_b/V_a \approx K_a C_a / \{C_b \times ([H^+]+K_a)\}$ (thin line). The latter is only close just before the equivalence point.

Clearly the above, piecemeal approach is flawed. It gets even more awkward and approximate when applied to polyprotic acids: for phosphoric acid, one would need four approximate segments plus four single points (the starting point of the titration, and the three equivalence points). For citric acid (a triprotic acid with pK_as that lie close together) even that method fails miserably. Moreover, it is not clear how this approach can be extended to mixtures.

In short, since the traditional approach is of limited applicability, is often quite inaccurate, and at any rate is more complicated than the exact results of Section 3.1, there seems to be no good reason why it should still be propagated or used. As Ricci wrote in the introduction to his 1952 book on *Hydrogen Ion Concentration, New Concepts in a Systematic Treatment*: "To derive approximate practical formulas by simplification of exact mathematical equations is a legitimate and safe procedure, but to derive practical formulas on the basis of approximate thinking is unsafe, unsound, and in this case unnecessary." It is regrettable that this advice has been ignored for so long, especially in a discipline that values precision and accuracy.

4 Buffers

The previous chapter emphasized the steep change in pH near the equivalence point of a successful titration. Here we will focus on its counterpart: the regions in the titration curve where the pH is responding very little to the addition of strong acid or base. In many chemical systems, and especially in living cells, maintaining the pH within narrow bounds is essential, because the functioning of enzymes (the catalysts of chemical processes in living tissue) depends strongly on pH. The parts of a titration curve that vary little as acid or base is added are called the *buffer regions*, and the corresponding pH-stabilization (even though it is passive in the sense that it does not consume metabolic energy) is called *buffer action*.

4.1 Buffer strength

It is possible to define buffer action in terms of a titration curve, but this has the disadvantage that the property so defined would depend on the properties of two solutions, the sample and the titrant. A more useful definition, which depends only on the properties of a given solution, is obtained when we assume that the titrant is an infinitely concentrated strong acid or base. Van Slyke (1922) therefore defined the *buffer value* β as

$$\beta = C_b\left(\frac{\mathrm{d}(V_b/V_a)}{\mathrm{d}(\mathrm{pH})}\right)_{C_b\to\infty} = -\ln(10)\times C_b\left(\frac{\mathrm{d}(V_b/V_a)}{\mathrm{d}(\ln[\mathrm{H}^+])}\right)_{C_b\to\infty}$$

$$= -\ln(10)\times C_b\,[\mathrm{H}^+]\left(\frac{\mathrm{d}(V_b/V_a)}{\mathrm{d}[\mathrm{H}^+]}\right)_{C_b\to\infty} \tag{4.1}$$

where the factor $\ln 10 \approx 2.3$ comes from the conversion from $\mathrm{pH} = -\log[\mathrm{H}^+]$ to $\ln[\mathrm{H}^+]$. Here we will use the closely related definition of the *buffer strength* B as

$$B = \frac{\beta}{\ln(10)} = -\,C_b\,[\mathrm{H}^+]\left(\frac{\mathrm{d}(V_b/V_a)}{\mathrm{d}[\mathrm{H}^+]}\right)_{C_b\to\infty} \tag{4.2}$$

which yields somewhat simpler formulas.

While the above formulas apply to the addition of base, a similar formula can be given for the effect of added acid, viz.

$$B = \frac{\beta}{\ln(10)} = +\,C_a\,[\mathrm{H}^+]\left(\frac{\mathrm{d}(V_a/V_b)}{\mathrm{d}[\mathrm{H}^+]}\right)_{C_a\to\infty} \tag{4.3}$$

For a strong acid or base, the combination of eqns (3.4) and (4.2), or of the inverse of eqn (3.4) with eqn (4.3), yields the very simple result

$$B = [\mathrm{H}^+] + [\mathrm{OH}^-] \tag{4.4}$$

so that, for a concentrated monoprotic acid or base, B is for all practical purposes equal to its concentration C.

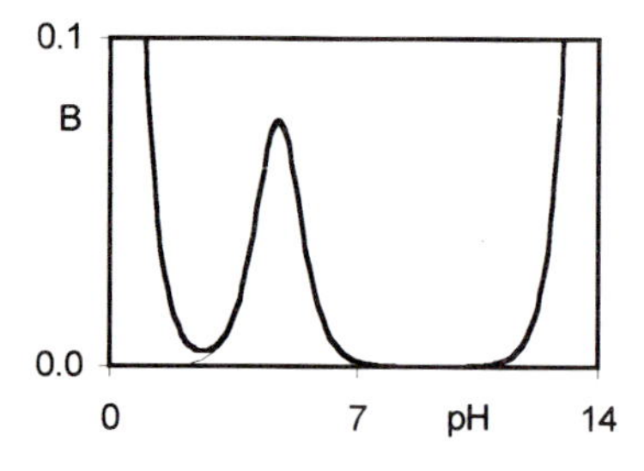

Fig. 4.1 The buffer strength B as a function of pH of acetic acid + acetate at a total analytical concentration C of 0.3 M. The thin line shows the contribution of the term $\alpha_1\alpha_0 C$.

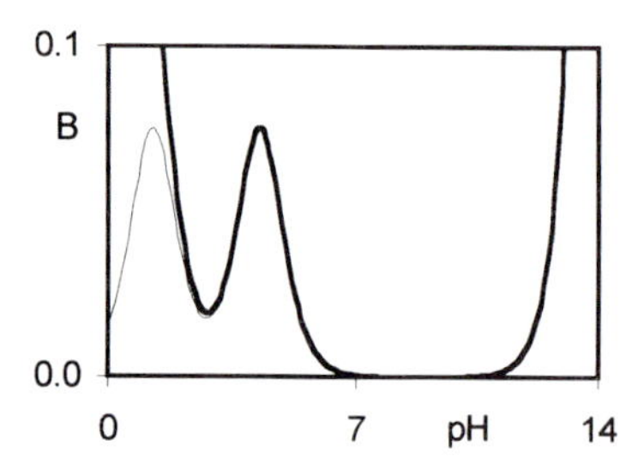

Fig. 4.2 The buffer strength B as a function of pH of oxalic acid and its salts at C = 0.3 M. The buffer region around pK_{a1} = 1.25 is largely hidden below the buffer strength of the strong acid used to obtain a low pH.

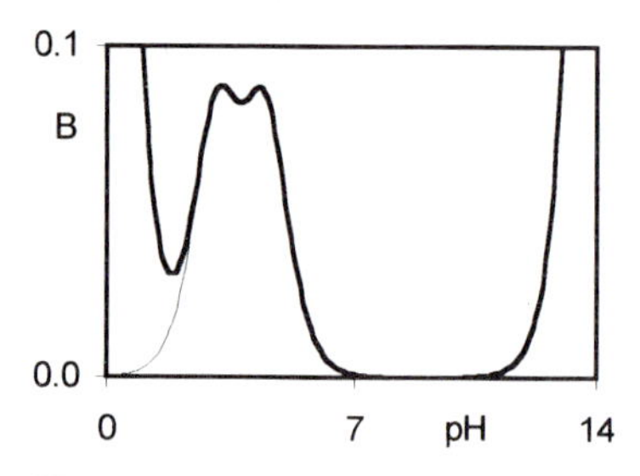

Fig. 4.3 The buffer strength B as a function of pH of fumaric acid (pK_{a1} = 3.05, pK_{a2} = 4.49) and its salts, C = 0.3 M. The two pK_as are close so that the buffer regions overlap.

For a single weak monoprotic acid or base of total analytical concentration C we likewise find

$$B = [\mathrm{H}^+] + \alpha_1\alpha_0 C + [\mathrm{OH}^-] \tag{4.5}$$

and for a mixture of two or more such weak monoprotic acids or bases

$$B = [\mathrm{H}^+] + \sum_i \alpha_{i1}\alpha_{i0} C_i + [\mathrm{OH}^-] \tag{4.6}$$

For a single diprotic weak acid or base we have

$$B = [\mathrm{H}^+] + (\alpha_2\alpha_1 + 4\alpha_2\alpha_0 + \alpha_1\alpha_0)C + [\mathrm{OH}^-] \tag{4.7}$$

and for a single weak triprotic acid or base

$$B = [\mathrm{H}^+] + (\alpha_3\alpha_2 + 4\alpha_3\alpha_1 + 9\alpha_3\alpha_0 + \alpha_2\alpha_1 + 4\alpha_2\alpha_0 + \alpha_1\alpha_0)C + [\mathrm{OH}^-] \tag{4.8}$$

and so on, where the coefficients are the squares of the index differences on the alphas: $4 = (3-1)^2 = (2-0)^2$, $9 = (3-0)^3$. Again, for mixtures of various non-interchangeable species, we merely add the various contributions as we did in eqn (4.6).

For an arbitrary mixture of i non-interchangeable components we have the general expression

$$B = [\mathrm{H}^+] + \sum_i C_i \sum_{j=1}^{j_{max}} \sum_{k=1}^{j} (j-k)^2 \alpha_{ij}\alpha_{ik} + [\mathrm{OH}^-] \tag{4.9}$$

Equation (4.9) can be used for *universal buffers*, i.e., mixtures of acids and their conjugated bases which provide nearly constant buffer strength over a wide range of pH values.

Figures 4.1 through 4.5 illustrate the application of the above relations.

4.2 The Henderson approximation

The above relations are exact in the sense that they follow directly from the definitions (4.2) or (4.3) of buffer strength B. Often, buffer solutions are formed by making solutions of an acid with its conjugated base, such as of acetic acid with sodium acetate, or of ammonium chloride with ammonia. In the derivation below we will consider a mixture of HAc and NaAc. We will first obtain an exact relation, due to Charlot (1951). Subsequently we will simplify this to a useful approximation found by L. J. Henderson (1908), so that we can see clearly what simplifying assumptions are involved.

Let the acid solution have the analytical concentrations C_{acid} and C_{base}, i.e., they were made by adding the amounts $C_{acid}V$ and $C_{base}V$ to a volume V of water. We then have the mass balance equations

$$[\mathrm{HAc}] + [\mathrm{Ac}^-] = C_{acid} + C_{base} \tag{4.10}$$

where the total analytical concentration C is equal to $C_{acid} + C_{base}$,

$$[\mathrm{Na}^+] = C_{base} \tag{4.11}$$

and the charge balance

$$[\mathrm{H}^+] + [\mathrm{Na}^+] = [\mathrm{Ac}^-] + [\mathrm{OH}^-] \tag{4.12}$$

We combine eqns (4.11) and (4.12) to

$$[\mathrm{Ac}^-] = C_{base} + \Delta \tag{4.13}$$

and substitute this into eqn (4.10) to obtain

$$[\mathrm{HAc}] = C_{acid} - \Delta \tag{4.14}$$

which can be combined with the definition (1.4) of the acid dissociation constant K_a to yield the *Charlot equation*

$$[\mathrm{H}^+] = \frac{K_a[\mathrm{HAc}]}{[\mathrm{Ac}^-]} = \frac{K_a(C_{acid} - \Delta)}{C_{base} + \Delta} \tag{4.15}$$

With the simplifying assumptions $C_{acid} \gg |\Delta|$ and $C_{base} \gg |\Delta|$ we can simplify the Charlot equation to the *Henderson approximation*

$$[\mathrm{H}^+] \approx \frac{K_a\, C_{acid}}{C_{base}} \tag{4.16}$$

Equation (4.16) has the same *form* as the expression (1.4) for the acid dissociation constant, but a quite different meaning, because C_{acid} and C_{base} are calculated on the basis of amounts of acid and base weighed rather than on actual concentrations of species. The differences between these follow directly from eqns (4.13) and (4.14) as $C_{acid} = [\mathrm{HAc}] + \Delta$, and $C_{base} = [\mathrm{Ac}^-] - \Delta$.

The assumptions $C_{acid} \gg |\Delta|$ and $C_{base} \gg |\Delta|$ are *both* applicable when $C = C_{acid} + C_{base} \gg |\Delta|$, i.e., when the system point lies well within the central triangle, *and* when the ratio C_{acid} / C_{base} is not too far from 1, i.e., when one restricts its application to the region around pH = pK_a. Moreover, there should not be a nearby pK_a-value, since the Charlot and Henderson expressions apply only to a single, monoprotic acid–base couple.

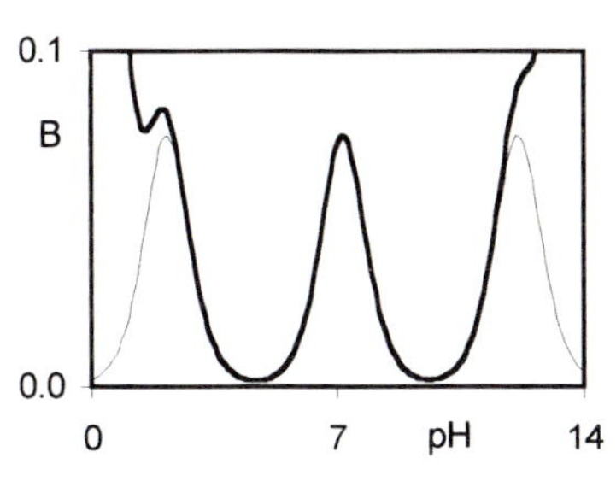

Fig. 4.4 The buffer strength *B* as a function of pH of phosphoric acid (pK_{a1} = 2.15, pK_{a2} = 7.20, pK_{a3} = 12.15) and its salts, *C* = 0.3 M. The three pK_as are well-separated, but two of the buffer regions overlap with those of the water.

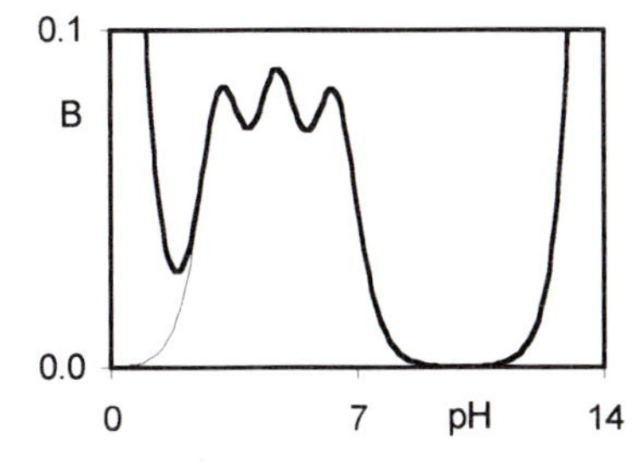

Fig. 4.5 The buffer strength *B* as a function of pH of citric acid (pK_{a1} = 3.13, pK_{a2} = 4.76, pK_{a3} = 6.40) and its salts, *C* = 0.3 M. The three pK_as are close, so that their buffer regions overlap.

Under conditions such that the Henderson approximation is applicable, a corresponding simplification of the expression for the buffer strength can also be found. We start from eqn (4.5) which, upon neglecting the terms in $[\mathrm{H}^+]$ and $[\mathrm{OH}^-]$, reduces to $B \approx \alpha_1\alpha_0\, C = \{[\mathrm{HAc}]\,[\mathrm{Ac}^-]\}/C$ because $\alpha_1 = [\mathrm{HAc}]/C$ and $\alpha_0 = [\mathrm{Ac}^-]/C$. Use of the above approximations $C_{acid} \approx [\mathrm{HAc}]$ and $C_{base} \approx [\mathrm{Ac}^-]$ then leads to

$$B \approx \frac{C_{acid}C_{base}}{C_{acid} + C_{base}} \tag{4.17}$$

or

$$\frac{1}{B} \approx \frac{1}{C_{acid}} + \frac{1}{C_{base}} \tag{4.18}$$

5 Other ionic equilibria

Many ionic equilibria can be described through formalisms similar to that used for acid–base phenomena. In this chapter we will briefly illustrate this, without going into great detail: once the analogy is established, the power and convenience of extending the formalism to these cases will be clear.

5.1 Complexation

The successive complexation of metal ions by ligands has many features in common with the stepwise protonation bases. But there is one annoying difference: the formation of metal–ligand complexes is typically described in terms of *formation constants* K_f rather than ligand dissociation constants K_d, where the latter would be the direct analogs of acid dissociation constants K_a. For a metal ion M forming a 1:1 complex with a ligand L we have

$$K_f = \frac{[\mathrm{ML}]}{[\mathrm{M}][\mathrm{L}]} \tag{5.1}$$

where we have left out the respective valencies. We can now define the concentration fractions α_1 and α_0 of metal ions with and without an attached ligand as

$$\alpha_0 = \frac{[\mathrm{M}]}{[\mathrm{M}]+[\mathrm{ML}]} = \frac{1}{1+K_f[\mathrm{L}]} \tag{5.2}$$

$$\alpha_1 = \frac{[\mathrm{ML}]}{[\mathrm{M}]+[\mathrm{ML}]} = \frac{K_f[\mathrm{L}]}{1+K_f[\mathrm{L}]} \tag{5.3}$$

which are analogous to eqns (1.22) and (1.23) except for the inversion of K_f.

The titration of such a metal ion M with such a ligand L is described by

$$\frac{V_L}{V_M} = \frac{\alpha_1 C_M + [\mathrm{L}]}{C_L - [\mathrm{L}]} \tag{5.4}$$

which is fully equivalent to the titration of a weak base with a strong acid. An analytically important group of predominantly 1:1 metal–ligand complexes are the metal chelates, and titrations based on their formation are common indeed, especially in the determination of water hardness.

Many metal ions can also form polyligand complexes with ligands such as ammonia or the halides. For example, for the successive formation of chloro complexes of cadmium(II), the formation constants are defined as

$$K_{f1} = \frac{[\mathrm{CdCl}^+]}{[\mathrm{Cd}^{2+}][\mathrm{Cl}^-]} \tag{5.5}$$

$$K_{f2} = \frac{[\mathrm{CdCl_2}]}{[\mathrm{CdCl^+}][\mathrm{Cl^-}]} \tag{5.6}$$

$$K_{f3} = \frac{[\mathrm{CdCl_3^-}]}{[\mathrm{CdCl_2}][\mathrm{Cl^-}]} \tag{5.7}$$

$$K_{f4} = \frac{[\mathrm{CdCl_4^{2-}}]}{[\mathrm{CdCl_3^-}][\mathrm{Cl^-}]} \tag{5.8}$$

The corresponding concentration fractions are

$$\alpha_4 = \frac{[\mathrm{CdCl_4^{2-}}]}{C} = \frac{K_{f1}K_{f2}K_{f3}K_{f4}[\mathrm{Cl^-}]^4}{K_{f1}K_{f2}K_{f3}K_{f4}[\mathrm{Cl^-}]^4 + K_{f1}K_{f2}K_{f3}[\mathrm{Cl^-}]^3 + K_{f1}K_{f2}[\mathrm{Cl^-}]^2 + K_{f1}[\mathrm{Cl^-}] + 1} \tag{5.9}$$

$$\alpha_3 = \frac{[\mathrm{CdCl_3^-}]}{C} = \frac{K_{f1}K_{f2}K_{f3}[\mathrm{Cl^-}]^3}{K_{f1}K_{f2}K_{f3}K_{f4}[\mathrm{Cl^-}]^4 + K_{f1}K_{f2}K_{f3}[\mathrm{Cl^-}]^3 + K_{f1}K_{f2}[\mathrm{Cl^-}]^2 + K_{f1}[\mathrm{Cl^-}] + 1} \tag{5.10}$$

$$\alpha_2 = \frac{[\mathrm{CdCl_2}]}{C} = \frac{K_{f1}K_{f2}[\mathrm{Cl^-}]^2}{K_{f1}K_{f2}K_{f3}K_{f4}[\mathrm{Cl^-}]^4 + K_{f1}K_{f2}K_{f3}[\mathrm{Cl^-}]^3 + K_{f1}K_{f2}[\mathrm{Cl^-}]^2 + K_{f1}[\mathrm{Cl^-}] + 1} \tag{5.11}$$

$$\alpha_1 = \frac{[\mathrm{CdCl^+}]}{C} = \frac{K_{f1}[\mathrm{Cl^-}]}{K_{f1}K_{f2}K_{f3}K_{f4}[\mathrm{Cl^-}]^4 + K_{f1}K_{f2}K_{f3}[\mathrm{Cl^-}]^3 + K_{f1}K_{f2}[\mathrm{Cl^-}]^2 + K_{f1}[\mathrm{Cl^-}] + 1} \tag{5.12}$$

$$\alpha_0 = \frac{[\mathrm{Cd^{2+}}]}{C} = \frac{1}{K_{f1}K_{f2}K_{f3}K_{f4}[\mathrm{Cl^-}]^4 + K_{f1}K_{f2}K_{f3}[\mathrm{Cl^-}]^3 + K_{f1}K_{f2}[\mathrm{Cl^-}]^2 + K_{f1}[\mathrm{Cl^-}] + 1} \tag{5.13}$$

In order to get an overview of such a set of sequential equilibria, we can plot the concentration fractions, as shown in Fig. 5.1, or construct a logarithmic concentration diagram, as in Fig. 5.2. The latter can be used to estimate the concentrations of all species as a function of the ligand concentration [L]. In both cases, we use pL = –log[L] instead of pH as the independent variable. The titration of such a metal ion with its ligand can be described by

$$\frac{V_L}{V_M} = \frac{F_c C_M + [\mathrm{L}]}{C_L - [\mathrm{L}]} \tag{5.14}$$

where

$$F_c = \alpha_1 + 2\alpha_2 + 3\alpha_3 + 4\alpha_4 \tag{5.15}$$

which is, again, fully analogous to the titration of a polyprotic base with a strong monoprotic acid. Often, however, such titrations are analytically not very useful because the formation constants are too similar in value.

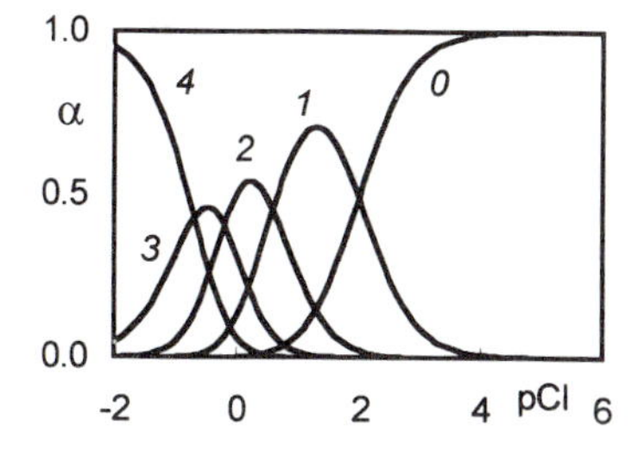

Fig. 5.1 The concentration fractions α for the various chloride complexes of Cd(II), computed with K_{f1} = 96 M^{-1}, K_{f2} = 4 M^{-1}, K_{f3} = 0.6 M^{-1}, K_{f4} = 0.2 M^{-1}. The labels indicate the number of chlorides bound to the central metal ion. For clarity the pCl scale has been extended to pCl = –2, even though an aqueous chloride concentration of 100 M is not realizable.

5.2 Extraction

Extraction is a process in which a given species is distributed over two contacting fluid phases (typically, two incompletely miscible solvents, such as water and an organic phase like carbon tetrachloride or toluene). The resulting equilibrium distribution of a species X over, say, an aqueous and a nonpolar organic phase, $X_{org} \rightleftharpoons X_{aq}$, is described by the partition coefficient K_p, for such a situation usually defined as

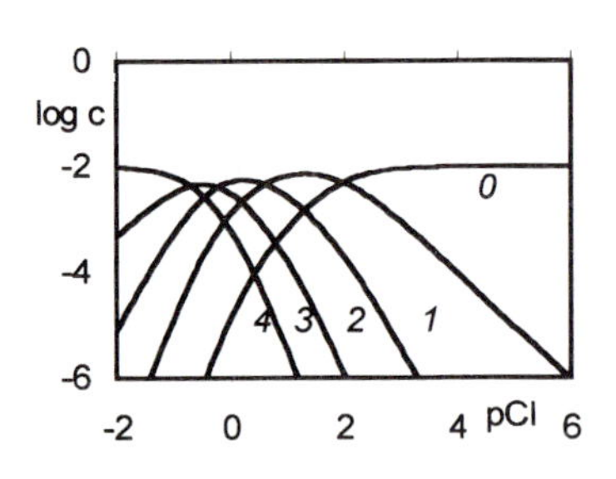

Fig. 5.2 The logarithmic concentration diagram for the various chloride complexes of Cd(II) at a total analytical cadmium concentration of 10 mM, for the data listed with Fig. 5.1.

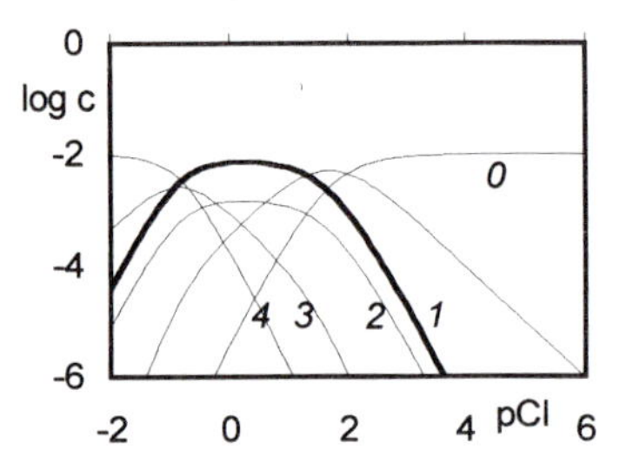

Fig. 5.3 The logarithmic concentration diagram for the various chloride complexes of Cd(II) at a total analytical cadmium concentration of 10 mM, in contact with an equal volume of organic phase with K_p = 5 for the uncharged species $CdCl_2$. Thin lines: aqueous concentrations; heavy line: $[CdCl_2]$ in organic phase.

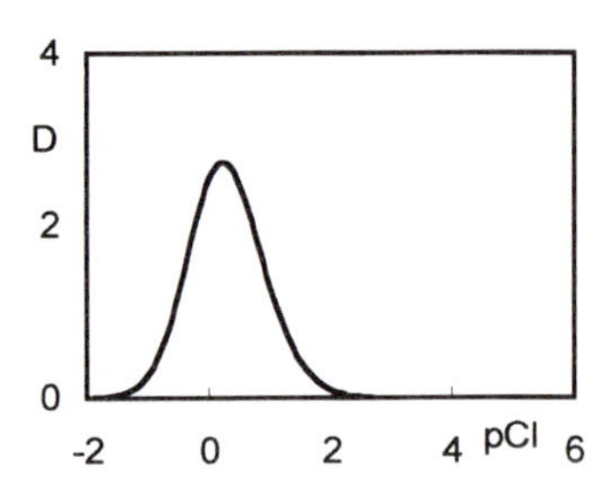

Fig. 5.4 The distribution $D = C_{org}/\,C_{aqueous}$ for the data of Fig. 5.3.

$$K_p = \frac{[\mathrm{X}]_{org}}{[\mathrm{X}]_{aq}} \tag{5.16}$$

The resulting *mass fractions* μ_{aq} and μ_{org} are then

$$\mu_{aq} = \frac{[\mathrm{X}]_{aq} V_{aq}}{[\mathrm{X}]_{aq} V_{aq} + [\mathrm{X}]_{aq} V_{aq}} \tag{5.17}$$

$$\mu_{org} = \frac{[\mathrm{X}]_{org} V_{org}}{[\mathrm{X}]_{aq} V_{aq} + [\mathrm{X}]_{aq} V_{aq}} \tag{5.18}$$

where V_{aq} and V_{org} denote the volumes of the aqueous and organic solvent respectively.

The extraction of metal ions often exploits their complexation equilibria, especially the much greater solubility in a non-polar organic phase of neutral over charged complexes. (This is the result of the substantial *Born energy* necessary to dissolve an ion rather than a corresponding neutral species in a solvent of low dielectric permittivity ε.) By adjusting the ligand concentration one can often manipulate the complexation equilibria of the metal of which extraction is desired, so that it is predominantly in the form of a neutral complex, while other metal ions are mostly in the form of charged complexes, which do not extract as effectively.

Figure 5.3 shows the logarithmic concentration diagram for the cadmium chloro complexes in water in contact with an equal volume of an organic phase with $K_p = 5$, and Fig. 5.4 shows the corresponding distribution of cadmium between the two phases, both as a function of pCl. In this case, maximal cadmium extraction is achieved by adjusting the pCl to about 1.5, i.e., for a free chloride concentration of about 0.03 M.

5.3 Solubility and precipitation

Many solubility problems are complicated by acid–base and complexation equilibria. For example, the sparingly soluble silver halides will dissolve in concentrated halide solutions through the formation of anionic complexes. Representing such equilibria in a logarithmic concentration diagram can provide a much needed overview of a complicated and otherwise almost impenetrable set of mathematical relations.

The chloro complexes of silver are described by the following equilibrium expressions:

$$AgCl_s \rightleftharpoons Ag^+ + Cl^- \qquad K_{s0} = [Ag^+]\,[Cl^-] \tag{5.19}$$

$$AgCl_s \rightleftharpoons AgCl \qquad K_{s1} = [AgCl] \tag{5.20}$$

$$AgCl_s + Cl^- \rightleftharpoons AgCl_2^- \qquad K_{s2} = [AgCl_2^-]\,/\,[Cl^-] \tag{5.21}$$

$$AgCl_s + 2Cl^- \rightleftharpoons AgCl_3^{2-} \qquad K_{s3} = [AgCl_3^{2-}]\,/\,[Cl^-]^2 \tag{5.22}$$

$$AgCl_s + 3Cl^- \rightleftharpoons AgCl_4^{3-} \qquad K_{s4} = [AgCl_4^{3-}]\,/\,[Cl^-]^3 \tag{5.23}$$

which are related to the formation constants through

$$K_{s1} = K_{s0}\, K_{f1} \tag{5.24}$$

$$K_{s2} = K_{s0}\, K_{f1}\, K_{f2} \tag{5.25}$$

$$K_{s3} = K_{s0}\, K_{f1}\, K_{f2}\, K_{f3} \tag{5.26}$$

$$K_{s4} = K_{s0}\, K_{f1}\, K_{f2}\, K_{f3}\, K_{f4} \tag{5.27}$$

The *solubility* S of silver is defined as the total analytical concentration of all soluble silver species, i.e., as

$$\begin{aligned} S &= [\mathrm{Ag}+] + [\mathrm{AgCl}] + [\mathrm{AgCl_2^-}] + [\mathrm{AgCl_3^{2-}}] + [\mathrm{AgCl_4^{3-}}] \\ &= K_{s0}\,\{1/[\mathrm{Cl^-}] + K_{f1} + K_{f1}K_{f2}[\mathrm{Cl^-}] + K_{f1}K_{f2}K_{f3}[\mathrm{Cl^-}]^2 + K_{f1}K_{f2}K_{f3}K_{f4}[\mathrm{Cl^-}]^3\} \\ &= K_{s0}/[\mathrm{Cl^-}] + K_{s1} + K_{s2}[\mathrm{Cl^-}] + K_{s3}[\mathrm{Cl^-}]^2 + K_{s4}[\mathrm{Cl^-}]^3 \end{aligned} \tag{5.28}$$

as long as the solution is in equilibrium with solid AgCl. The resulting logarithmic concentration diagram contains only straight lines, see Fig. 5.5.

As soon as the product of solubility S times solution volume V exceeds the total amount of silver available, the solid will dissolve, and the system will revert to the usual situation in which the total analytical concentration C controls the speciation, as illustrated in Fig. 5.6.

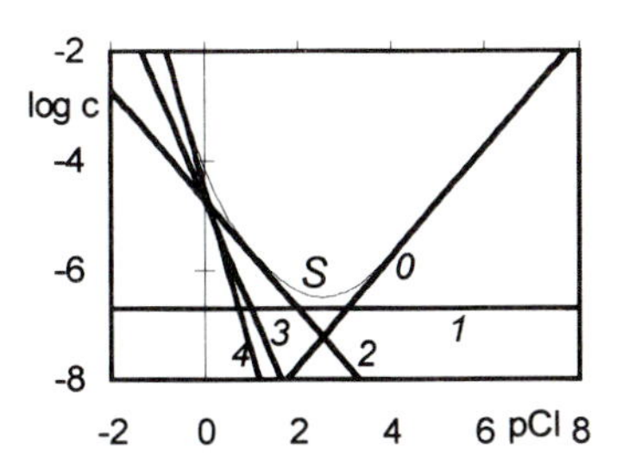

Fig. 5.5 The logarithmic concentration diagram for the silver chloride system in the presence of excess AgCl. The thin line labeled S shows the corresponding solubility. Equilibrium constants used: pK_{s0} = 9.75, pK_{s1} = 6.05, pK_{s2} = 4.13, pK_{s3} = 3.3, pK_{s4} = 3.6.

5.4 Precipitation titrations

Argentometry is the use of silver ions to titrate halides and pseudo-halides. Such titrations are readily described in terms of the formalism of Chapter 3, as demonstrated here in a simplified treatment that only takes into account the dominant silver species in such a titration before the equivalence point, Ag^+ and AgCl. (Including the higher silver complexes is quite straightforward, but is not included here as it merely clutters up the formalism, whereas we want to focus here on the general features of the approach.)

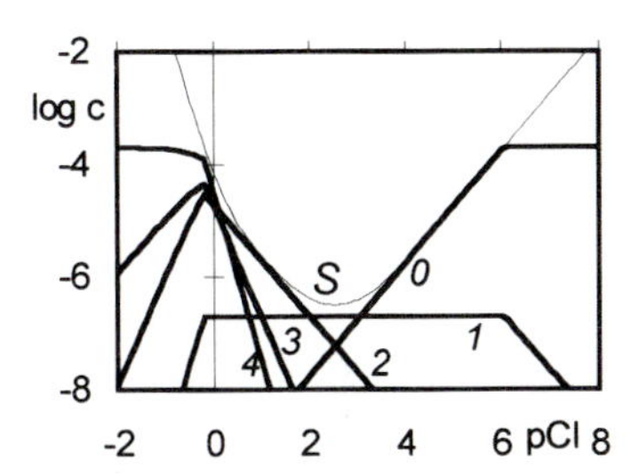

Fig. 5.6 The diagram of Fig. 5.5 when the amount of silver available is limited to a total analytical concentration of 0.2 mM, so that the precipitate dissolves as soon as S exceeds 0.2 mM.

We consider the titration of a sample of volume V_s containing chloride at a concentration C_s with a titrant containing C_t M of some silver salt, such as $AgNO_3$ or AgAc. The mass balance relations are now

$$[\mathrm{Cl^-}]+[\mathrm{AgCl}]+P_{\mathrm{AgCl}} = \frac{C_sV_s}{V_s+V_t} \tag{5.29}$$

$$[\mathrm{Ag^+}]+[\mathrm{AgCl}]+P_{\mathrm{AgCl}} = \frac{C_tV_t}{V_s+V_t} \tag{5.30}$$

where P_{AgCl} represents the concentration of AgCl removed by precipitation. Subtraction yields

$$[\mathrm{Cl^-}]-[\mathrm{Ag^+}] = \frac{C_sV_s - C_tV_t}{V_s+V_t} \tag{5.31}$$

which can be rewritten as

$$V_t\{C_t+[\mathrm{Cl^-}]-[\mathrm{Ag^+}]\} = V_s\{C_s-[\mathrm{Cl^-}]+[\mathrm{Ag^+}]\} \tag{5.32}$$

or

$$\frac{V_t}{V_s} = \frac{C_s+[\mathrm{Ag^+}]-[\mathrm{Cl^-}]}{C_t-[\mathrm{Ag^+}]+[\mathrm{Cl^-}]} = \frac{C_s+[\mathrm{Ag^+}]-K_{s0}/[\mathrm{Ag^+}]}{C_t-[\mathrm{Ag^+}]+K_{s0}/[\mathrm{Ag^+}]} \tag{5.33}$$

which will be recognized as fully equivalent to the expression for the progress of the titration of a strong base with a strong acid,

$$\frac{V_a}{V_b}=\frac{C_b+[\mathrm{H}^+]-[\mathrm{OH}^-]}{C_a-[\mathrm{H}^+]+[\mathrm{OH}^-]}=\frac{C_s+[\mathrm{H}^+]-K_w/[\mathrm{H}^+]}{C_t-[\mathrm{H}^+]+K_w/[\mathrm{H}^+]} \tag{5.34}$$

where K_{s0} has taken the place of K_w. The analogy with acid–base titrations also applies to Gran plots, which in this case are quite satisfactory, and allow a precise linear least-squares extrapolation to the equivalence point. Alternatively one can fit the entire curve with a non-linear least squares routine.

Because the silver salts of chloride, bromide, and iodide have quite different K_{s0}-values, it is possible to titrate all three simultaneously, in one titration, see Fig. 5.7. Even in this case, Gran plots are quite efficient in finding the equivalence points, as can be seen in Fig. 5.8.

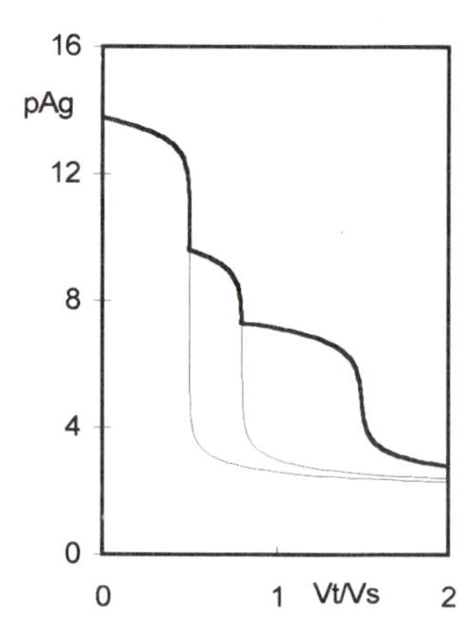

Fig. 5.7 Titration curves for the titration of 5 mM NaI + 3 mM NaBr + 7 mM NaCl with 10 mM $AgNO_3$. The thin lines show the titration of 5 mM NaI, and that of 5 mM NaI + 3 mM NaBr. Equilibrium constants used: $pK_{s0,AgI}$ = 16.08, $pK_{s0,AgBr}$ = 12.30, and $pK_{s0,AgCl}$ = 9.75.

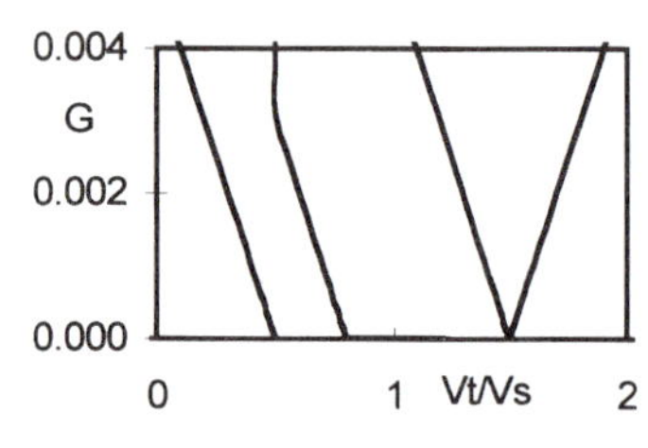

Fig. 5.8 Gran plots for the titration of 5 mM NaI + 3 mM NaBr + 7 mM NaCl with 10 mM $AgNO_3$. The Gran function G is given by, from left to right, (V_t/V_s+1) times $K_{s0,AgI}/[Ag^+]$, $K_{s0,AgBr}/[Ag^+]$, $K_{s0,AgCl}/[Ag^+]$, and $[Ag^+]$ respectively, where $[Ag^+]$ is obtained from the measured value of the potential E, via $\log[Ag^+] = (E - E^o_{Ag^+/Ag})/0.059$.

5.5 Reduction and oxidation

Redox equilibria are described by the *Nernst equation.* For a one-electron transfer between an oxidized form O and its reduced counterpart R the Nernst equation takes the form

$$fE = fE^o_{\mathrm{OR}} + \log\frac{[\mathrm{O}]}{[\mathrm{R}]} \tag{5.35}$$

where f = F/{RT log(10)} or about 16.9 V^{-1} at 25°C; under those circumstances, $1/f$ = 0.059 V. Equation (5.35) is analogous to the equilibrium expression for proton transfer,

$$\mathrm{pH} = \mathrm{p}K_a + \log\frac{[\mathrm{A}^-]}{[\mathrm{HA}]} \tag{5.36}$$

We will exploit this analogy by introducing the abbreviations

$$h = 10^{-fE}, \qquad k = 10^{-fE^o_{\mathrm{OR}}} \tag{5.37}$$

which allow us to rewrite eqn (5.35) as

$$\mathrm{p}h = \mathrm{p}k + \log\frac{[\mathrm{O}]}{[\mathrm{R}]} \tag{5.38}$$

As will be illustrated below, this formalism, which is readily extended to multi-electron transfers, allows us to write down expressions for redox equilibria merely by analogy to the corresponding acid–base expressions.

Consider the redox equilibrium $Fe^{3+} + e^- \rightleftharpoons Fe^{2+}$. When the total analytical concentration is C, the concentration fractions of Fe^{3+} and Fe^{2+} are

$$\alpha_{\mathrm{Fe}^{2+}} = \frac{h}{h+k}, \qquad \alpha_{\mathrm{Fe}^{3+}} = \frac{k}{h+k} \tag{5.39}$$

which can be compared with

$$\alpha_{\mathrm{HA}} = \frac{[\mathrm{H}^+]}{[\mathrm{H}^+]+K_a}, \qquad \alpha_{\mathrm{A}^-} = \frac{K_a}{[\mathrm{H}^+]+K_a} \tag{5.40}$$

We can then construct logarithmic concentration diagrams that show the various species as a function of $\mathrm{p}h = f\,E = E/0.059$ (at room temperature), and compute the potential E of the solution using the *proton condition*, which counts electron gainers and losers in the same way as the proton condition does for protons.

In order to describe the *redox titration* of, say, ferrous ammonium sulfate with ceric sulfate in sulfuric acid, we will here assume that the sample and titrant initially contain only Fe^{2+} and Ce^{4+} respectively. The mass balance relations for the various species now read

$$[Fe^{2+}] = \frac{h}{h+k_{Fe}} \frac{C_s V_t}{V_s+V_t}, \qquad [Fe^{3+}] = \frac{k_{Fe}}{h+k_{Fe}} \frac{C_s V_t}{V_s+V_t} \tag{5.41}$$

$$[Ce^{3+}] = \frac{h}{h+k_{Ce}} \frac{C_t V_t}{V_s+V_t}, \qquad [Ce^{4+}] = \frac{k_{Ce}}{h+k_{Ce}} \frac{C_t V_t}{V_s+V_t} \tag{5.42}$$

$$h = 10^{-fE}, \qquad k_{Fe} = 10^{-fE^0_{Fe32}}, \qquad k_{Ce} = 10^{-fE^0_{Ce43}} \tag{5.43}$$

Finally, during the titration we have the electron condition

$$[Fe^{3+}] = [Ce^{3+}] \tag{5.44}$$

because the electrons lost in the oxidation of Fe^{2+} to Fe^{3+} were gained by Ce^{4+} in its corresponding reduction to Ce^{3+}. Substitution of eqns (5.41) and (5.42) into (5.44) then yields

$$\frac{V_t}{V_s} = \frac{C_s\, k_{Fe}(h+k_{Ce})}{C_t\,(h+k_{Fe})\,h} = \frac{C_s \alpha_{Fe^{3+}}}{C_t \alpha_{Ce^{3+}}} \tag{5.45}$$

which is the sought equation for the progress of the titration. Note that this is equivalent to the titration of a weak acid with a weak base, or vice versa, except that the terms in Δ of the acid–base progress curve are missing. This is because the reduction or oxidation of the solvent (equivalent to the water autodissociation terms $\pm\Delta$ in the acid–base case) usually need not be taken into account, and seldom can be described in terms of equilibria anyway. As a consequence, Gran plots are applicable, and can be used to find the equivalence point, as a valid alternative to fitting the entire titration curve.

Just as the description of the progress of a titration of weak acid with strong base could be expanded to encompass all acid–base titrations, the above formalism can be extended to all redox titrations that have sufficiently fast kinetics to obey equilibrium relations (de Levie 1992b). For example, the titration of ferrous ion with permanganate in acidic solution is described by

$$\frac{V_t}{V_s} = \frac{C_s \alpha_{Fe^{3+}}}{5\, C_t \alpha_{Mn^{2+}}} = \frac{C_s k_{Fe}(h^5 + k^5_{Mn})}{5\, C_t\,(h+k_{Fe})\, h^5} \tag{5.46}$$

where the coefficient 5 reflects the five-electron transfer in the reduction of MnO_4^- to Mn^{2+}. Figure 5.9 illustrates this titration curve, and Fig. 5.10 the corresponding Gran plots.

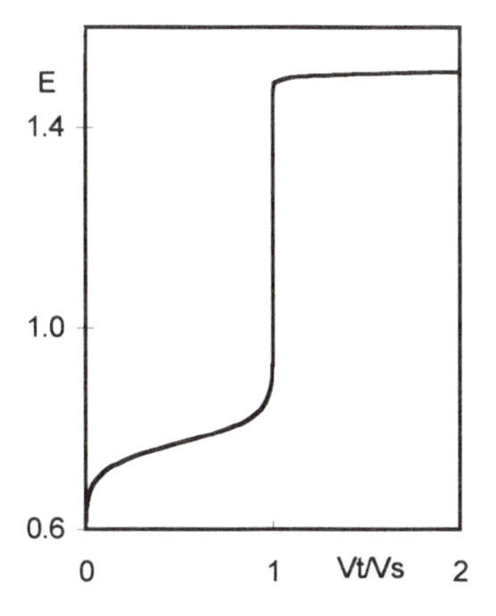

Fig. 5.9 Titration curves for the titration of Fe(II) with MnO_4^- at pH = 0. Equilibrium constants used: E^0_{Fe32} = 0.771 V, E^0_{Mn72} = 1.51 V.

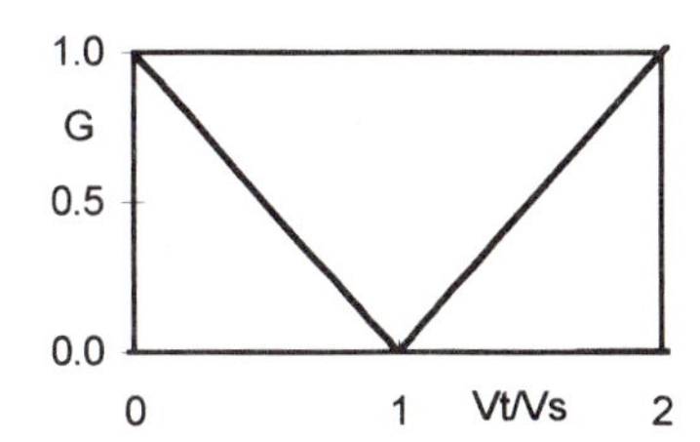

Fig. 5.10 Gran plots for the titration of Fe(II) with MnO_4^- at pH = 0. The Gran function G is $hV_t/k_s V_s$ (left) or $(k_t/h)^5$ (right).

The redox potential is a solution property much like its pH, and it can be stabilized by buffer action. The redox buffer strength of biological systems can be as important to its proper functioning as its acid–base buffer strength. Again, the formalism is strikingly similar to that of the acid–base case.

For example, a solution containing both ferrous and ferric ions has a redox buffer strength (de Levie 1999a)

$$\frac{1}{B} = \frac{1}{[Fe^{3+}]} + \frac{1}{[Fe^{2+}]} \tag{5.47}$$

while the buffer strength of a solution containing both MnO_4^- and Mn^{2+} is

$$\frac{5^2}{B} = \frac{1}{[MnO_4^-]} + \frac{1}{[Mn^{2+}]} \tag{5.48}$$

where 5 reflects the number of electrons transferred in going from MnO_4^- to Mn^{2+}.

6 Activity effects

So far we have treated equilibrium parameters such as K_a, K_w, K_f, K_p, K_s, and E^o as true constants. They will, of course, depend on temperature and (at least in principle, though in essentially non-compressible solutions not much) on pressure. But at constant temperature and pressure, the equilibrium *constants* should be just that: constants. However, in practice, they are only *approximately* constant. This chapter explains why this is so, and what can be done to make them more constant. That remedy works well in dilute solutions, but is only partially successful in concentrated solutions, for practical as well as theoretical reasons that are beyond the scope of this primer.

6.1 Why activity effects?

The experimental evidence from almost a century of precise measurements of equilibrium constants shows that, in dilute solutions, they are essentially constant only when compared at constant *ionic strength*, a parameter introduced empirically by Lewis and Randall (1921) and soon thereafter firmly anchored in theory by Debye and Hückel (1923).

The ionic strength I is a measure which counts the ions in solution, with each ionic concentration c_i weighed by the square of its valence z_i,

$$I = \frac{1}{2}\sum_i z_i^2 c_i \tag{6.1}$$

where the factor ½ serves to make the ionic strength equal to the concentration for the solution of a single, strong 1:1 electrolyte: a 0.37 M NaCl solution has an ionic strength of 0.37 M. By excluding the contributions from uncharged species, the ionic strength clearly reflects *coulombic* interactions. The way these affect ionic equilibria can be envisioned as follows.

Electroneutrality requires that a macroscopic volume of an electrolyte solution contain equivalent amounts anions and cations. The individual ions can move relatively freely throughout the solution, but of course anions and cations attract each other, while they repel ions of the same charge type. As a result, the average anion–cation distance is smaller than the average anion–anion and cation–cation distances, so that on average the ions are coulombically attracted (and therefore have a lower energy) by being in the electrolyte solution, the more so the higher its ionic strength. This lowered energy causes the (weak) dependence of the equilibrium constants on ionic strength.

The effect can be treated through a correction term, the *activity coefficient* f, which can be introduced into the expressions for equilibrium constants as a multiplier of all concentrations c. The resulting product $f\,c$ is called the *activity*, and is often denoted by the symbol a.

It is important to realize where activity corrections apply, and where they do not. Since interionic attractions can lower the energetics of ions, they can affect the numerical values of the equilibrium constants. However, activity effects have no direct influence on the mass and charge balance equations, which are purely accounting methods, or on any relations derived from these, such as the proton and electron conditions. Similarly, activity effects do not change the equivalence volume V_e in a titration, but can change the shape of the titration curve. Fortunately, as we will see below, those changes are usually relatively minor.

6.2 Ionic interactions

Consider one liter of a 1 M solution of NaCl. It contains N_A sodium and chloride ions, where N_A is Avogadro's number, about 6×10^{23} mole^{-1}. Since no closed-form solution is known for three-body problems, a 10^{24}-body problem is clearly beyond reach, and we must make simplifying assumptions to analyze the problem. On the other hand, we are not really interested in the specific solutions for all anions and cations, but only in its statistical average. Debye and Hückel reduced the problem to a manageable form by considering the effect, on an arbitrarily selected 'central' ion, of all the other ions in solution as that of a smeared-out charge distribution with a total charge opposite to that of the central ion. The problem then becomes one of a central ion and its 'ionic cloud', a two-body problem. As you will see below, even such a drastic simplification already leads to rather complicated mathematics.

Here we will consider a symmetrical electrolyte such as NaCl or $MgSO_4$, i.e., with $z_+ = -z_- = |z|$. The concentrations c_+ and c_- of the anions and cations in the ionic cloud are then determined by the Boltzmann distribution,

$$c_+ = c^* \exp\left[\frac{-|z|F\psi}{RT}\right] \tag{6.2}$$

$$c_- = c^* \exp\left[\frac{+|z|F\psi}{RT}\right] \tag{6.3}$$

where F is the Faraday, R the gas constant, T the absolute temperature, ψ is the distance-dependent potential around the central ion, and c^* is the ionic concentration sufficiently far away from the central ion, where the concentration has its 'bulk' value, and the potential ψ is taken to be zero.

We now use the Poisson equation to link the ionic concentrations c to the potential ψ. Since we consider the ionic charges around a central ion, this requires spherical coordinates, i.e.,

$$\frac{1}{r^2}\frac{\mathrm{d}}{\mathrm{d}r}\left(r^2\frac{\mathrm{d}\psi}{\mathrm{d}r}\right) = -\frac{\rho}{\varepsilon} \tag{6.4}$$

where $\rho = |z|\ F\,(c_+ - c_-)$ is the charge density, r is the radial distance from the center of the central ion, and ε is the dielectric permittivity, i.e., the product of the (dimensionless) relative permittivity ε_{rel} and the vacuum permittivity ε_0 (of about 8.854 pF m^{-1}).

Substituting eqns (6.2) and (6.3) into (6.4) yields

$$\frac{1}{r^2}\frac{\mathrm{d}}{\mathrm{d}r}\left(r^2\frac{\mathrm{d}\psi}{\mathrm{d}r}\right)=\frac{|z|Fc^*}{\varepsilon}\left\{\exp\left[\frac{+|z|F\psi}{RT}\right]-\exp\left[\frac{-|z|F\psi}{RT}\right]\right\} \tag{6.5}$$

and introduction of the dimensionless parameters $y = |z|F\psi\,/RT$ and $\kappa^2 = z^2F^2c^*/\varepsilon RT$ leads to

$$\frac{1}{r^2}\frac{\mathrm{d}}{\mathrm{d}r}\left(r^2\frac{\mathrm{d}y}{\mathrm{d}r}\right)=\kappa^2\left(e^{+y}-e^{-y}\right) \tag{6.6}$$

Equation (6.6) has the general solution

$$\psi=\frac{A_-e^{-\kappa r}}{r}+\frac{A_+e^{+\kappa r}}{r} \tag{6.7}$$

which need not be pursued here. Instead we will merely consider the quantity $1/\kappa = \sqrt{(\varepsilon RT/z^2F^2c^*)}$ which is the *characteristic distance* in eqn (6.7). This quantity also plays a role in, e.g., the electrical double layer around colloid particles, and had already been studied by Gouy (1909, 1910).

Equations (6.6) and (6.7) do not involve the chemical nature of the ions, but merely their valencies, and can therefore be extended to electrolyte mixtures (Gouy 1910, Levine and Jones 1969), in which case we obtain

$$\kappa=\frac{F}{\sqrt{\varepsilon RT}}\sqrt{\sum_i z_i^2c_i^*}=\frac{F\sqrt{2}}{\sqrt{\varepsilon RT}}\sqrt{I} \tag{6.8}$$

where I is as defined in eqn (6.1). This is how the ionic strength I enters the problem.

The Debye–Hückel theory provides an explicit expression for the activity coefficient of an ion as a function of two parameters: the ionic strength I, and the distance of closest approach d of the central ion and its counter-ions (which parameter enters the model as an integration constant),

$$\log f_i=-\frac{z_i^2F^2}{8\pi\varepsilon N_ART}\frac{\kappa}{1+\kappa d}=-\frac{z_i^2A\sqrt{I}}{1+Bd\sqrt{I}} \tag{6.9}$$

where the constants have the numerical values $A = 0.51\ \mathrm{M}^{-1/2}$ and $B = 3.3\times10^9\ \mathrm{M}^{-1/2}\ \mathrm{m}^{-1}$ in water at 25°C. The distance of closest approach d depends on the radii of at least two species (and is therefore poorly defined in an electrolyte mixture), and has only a relatively minor effect on the activity coefficient. It is therefore convenient to replace it by the order-of-magnitude estimate d ≈ 3×10^{-10} m, in which case eqn (6.9) reduces to

$$\log f_i\approx-\frac{0.5\,z_i^2\sqrt{I}}{1+\sqrt{I}} \tag{6.10}$$

This expression is sometimes further simplified for use at very low ionic strengths, where $\sqrt{I} \ll 1$, in which case we obtain the so-called Debye–Hückel limiting law

$$\log f_i\approx-0.5\,z_i^2\sqrt{I} \tag{6.11}$$

6.3 Non-ionic interactions

So far we have only considered coulombic interactions, and these are indeed the dominant effect in dilute electrolyte solutions. However, neutral species in solutions can also be affected by the presence of ions in solution, as evidenced by so-called *salting-in* and *salting-out* effects of ions on the solubility of neutral species. Such effects often work through modifying the structure of water, or by tying up water as water of hydration. Since the same effects can be expected to work on ions, they are often included empirically through

$$\log f_i \approx -0.5\, z_i^2 \left(\frac{\sqrt{I}}{1+\sqrt{I}} - C\, I \right) \tag{6.12}$$

where the constant C depends on the nature of the ion i and on that of the other ions present. Davies (1938, 1962) has studied a large number of electrolyte solutions, and found an average value for C of 0.3. This, then, is the form in which the *Davies equation* is most often used, because it contains no adjustable parameters, yet represents the general trend (though not the specifics) of the activity coefficients of many ions in solution:

$$\log f_i \approx -0.5\, z_i^2 \left(\frac{\sqrt{I}}{1+\sqrt{I}} - 0.3\, I \right) \tag{6.13}$$

The Davies equation is a simplification of the Debye–Hückel equation (for $Bd = 1$) with a first-order correction for salting effects. It does not address the effects of neutral species on ions, or that of ions on neutral species. In concentrated solutions, all such effects play a role, and relations such as eqn (6.13) must be considered as first approximations only. Below we will use them as such.

6.4 Activity corrections

In order to use eqn (6.13) to make activity corrections, it is convenient to separate the ionic valence z_i by writing

$$\log f \approx -0.5 \left(\frac{\sqrt{I}}{1+\sqrt{I}} - 0.3\, I \right) \tag{6.14}$$

or

$$f = 10^{-0.5\left(\sqrt{I}/(1+\sqrt{I})-0.3I\right)} \tag{6.15}$$

and

$$\log f_i = z_i^2 \log f \qquad \text{or} \qquad f_i = f^{z_i^2} \tag{6.16}$$

because this leads to rather simple expressions for the activity corrections. Note that, at this level of approximation, neutral species (with $z = 0$) are assumed to require no activity corrections.

Now consider the equilibrium constants K_w, K_a, K_{a1}, and K_{a2} for water, for a weak monoprotic acid HA, and for a diprotic acid H_2A respectively. We have

$$K_w^t = [H^+]f_+[OH^-]f_- = [H^+]f[OH^-]f = K_w f^2 \tag{6.17}$$

$$K_a^t = \frac{[\mathrm{H}^+]f_+[\mathrm{A}^-]f_-}{[\mathrm{HA}]} = \frac{[\mathrm{H}^+]f[\mathrm{A}^-]f}{[\mathrm{HA}]} = K_a f^2 \tag{6.18}$$

$$K_{a1}^t = \frac{[\mathrm{H}^+]f_+[\mathrm{HA}^-]f_-}{[\mathrm{H_2A}]} = K_a f^2 \tag{6.19}$$

$$K_{a2}^t = \frac{[\mathrm{H}^+]f_+[\mathrm{A}^{2-}]f_{2-}}{[\mathrm{HA}^-]f_-} = \frac{[\mathrm{H}^+]f[\mathrm{A}^{2-}]f^4}{[\mathrm{HA}^-]f} = K_a f^4 \tag{6.20}$$

and so on, where the superscript t indicates the '*t*rue', *t*hermodynamic value (which of course is true only insofar as the activity correction is correct). However, one should be careful to consider the valencies of the species in the equilibrium expressions. For example, for the ammonium/ammonia monoprotic acid–base couple, $NH_4^+ \rightleftharpoons H^+ + NH_3$, we have

$$K_a^t = \frac{[\mathrm{H}^+]f_+[\mathrm{NH_3}]}{[\mathrm{NH_4^+}]f_+} = \frac{[\mathrm{H}^+]f[\mathrm{NH_3}]}{[\mathrm{NH_4^+}]f} = K_a \tag{6.21}$$

and for a diprotic aminoacid such as glycine, NH_3^+-CH_2-COO^- (here abbreviated to HGly), we find

$$K_{a1}^t = \frac{[\mathrm{H}^+]f_+[\mathrm{HGly}]}{[\mathrm{H_2Gly}^+]f_+} = K_{a1} \tag{6.22}$$

$$K_{a2}^t = \frac{[\mathrm{H}^+]f_+[\mathrm{Gly}^-]f_-}{[\mathrm{HGly}]} = K_a f^2 \tag{6.23}$$

We already commented on the fact that activity corrections have no place in balance equations, such as those accounting for mass and charge, or in the related proton or electron conditions. On the other hand, there are two places where activity corrections may be required. The first of these is in the values of the equilibrium constants, as illustrated in eqns (6.17) through (6.23). These relations must often be used in reverse, because most values for equilibrium constants listed in compilations are K^t-values, i.e., they have been corrected for activity effects or, equivalently, extrapolated to infinite dilution (where $f = 1$).

The second place where an activity correction is required is in *electrometric measurements*, such as with a pH meter or in recording a cell potential. Equilibrium electrometric measurements yield data that approximate the activities rather than the concentrations of the monitored species. In other words, the Nernst equation should be written in terms of activities rather than concentrations, and the measured pH_m must be interpreted as

$$\mathrm{pH}_m = -\log\left([\mathrm{H}^+]f_+\right) = \mathrm{pH} - \log f \tag{6.24}$$

where we retain the symbol pH for $-\log[H^+]$ just as we retain the notation of the equilibrium constants K_w, K_a, K_f, etc. for the concentration-based constants. Note that optical measurements of the pH, as in a photometric titration, should *not* be corrected for activities, because the concentrations in Beer's law are indeed concentrations, not activities.

6.5 An example

The titration of citric acid with NaOH provides a clear example of the magnitude of activity corrections. We will consider the following two simple scenarios: either citric acid and sodium hydroxide are the only electrolytes present, or the titration is performed in a solution containing a large excess of non-participating ('inert') ions, so that the ionic strength can be considered to be constant during the titration. We will treat the latter case first, because it is easier. The computations are readily performed with the help of a calculator or computer, and are especially easy on a spreadsheet.

When the ionic strength I is kept constant, we first compute $\log f$ from eqn (6.14) given the value of I, and then use this to calculate the values of pK_w, pK_{a1}, pK_{a2}, and pK_{a3} from the (supposedly infinite-dilution) literature values pK'_w, pK'_{a1}, pK'_{a2}, and pK'_{a3} using $pK_w = pK'_w + 2\log f$, $pK_{a1} = pK'_{a1} + 2\log f$, $pK_{a2} = pK'_{a2} + 4\log f$, $pK_{a3} = pK'_{a3} + 6\log f$. Table 6.1 shows the values for $\log f$ and for these four equilibrium constants at three different ionic strengths. The value of pH_m follows from eqn (6.24).

Table 6.1. The value of log *f* as calculated from eqn (6.14), of pK_w, and of the various pK_as of citric acid, all as a function of the ionic strength *I*.

I	$\log f$	pK_w	pK_{a1}	pK_{a2}	pK_{a3}
0	0	14.00	3.13	4.76	6.40
0.025	–0.065	13.87	3.00	4.50	6.01
0.1	–0.105	13.79	2.92	4.34	5.77
0.4	–0.134	13.73	2.86	4.23	5.60

With these new constants we can now compute activity-corrected titration curves. We see that the effect of ionic strength on pK_w and pK_{a1} is relatively minor, but that it is considerably larger on pK_{a3}, which involves the trivalent ion Cit^{3-}. Figure 6.1 illustrates the titration curves for the titration of 1 mM citric acid with 1 mM NaOH at these different ionic strengths.

When the titration is performed in the absence of additional electrolyte, then the ionic strength varies during the titration. In that case we must compute the activity correction *by iteration*, i.e., we first estimate I on the basis of uncorrected equilibrium constants through

$$I = 0.5([H^+]+[Na^+]+[OH^-]+[H_2Cit^+]+4[HCit^{2+}]+9[HCit^{3+}])$$

$$= 0.5([H^+]+\frac{C_bV_b/V_a}{1+V_b/V_a}+K_w/[H^+]$$

$$+\frac{[H^+]^2K_{a1}+4[H^+]K_{a1}K_{a2}+9K_{a1}K_{a2}K_{a3}}{[H^+]^3+[H^+]^2K_{a1}+[H^+]K_{a1}K_{a2}+K_{a1}K_{a2}K_{a3}}\frac{C_a}{1+V_b/V_a}) \qquad (6.25)$$

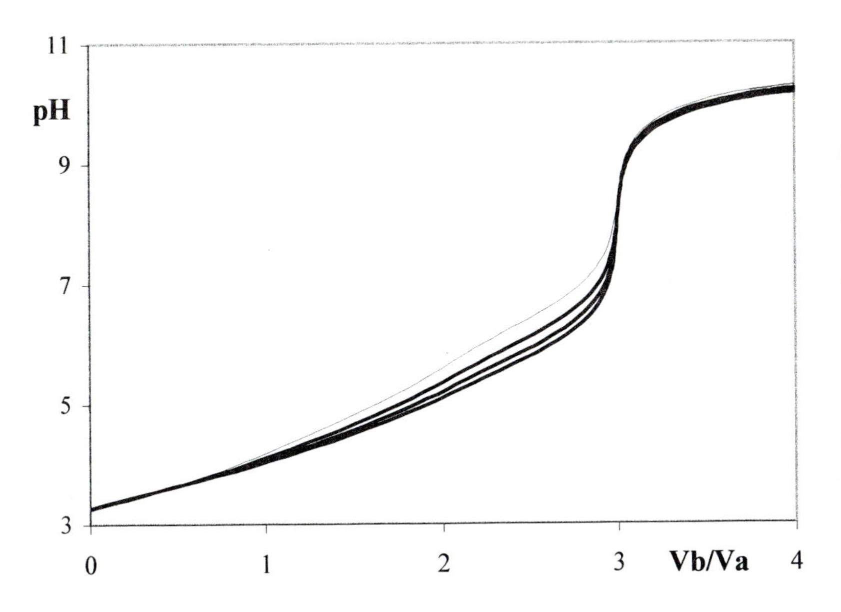

Fig. 6.1 Titration curves for the titration of 1 mM citric acid with 1 mM NaOH at various values of the ionic strength, calculated for $pK_w = 14.00$, $pK_{a1} = 3.13$, $pK_{a2} = 4.76$, $pK_{a3} = 6.40$, $C_a = C_b = 0.001$, and eqns (6.13) and (6.24). Heavy lines, from top to bottom, for $I = 0.025$ M, 0.1 M, and 0.4 M respectively. The thin line shows the curve calculated without any activity corrections.

With the estimated values of I we now correct the equilibrium constants, then use these for an improved estimate of I, etc. Although this sounds rather complicated, one seldom need more than one iteration cycle to find answers within the precision of the various pK-values, of about ±0.01. Figure 6.2 displays the results of such a calculation. In this example, the ionic strength

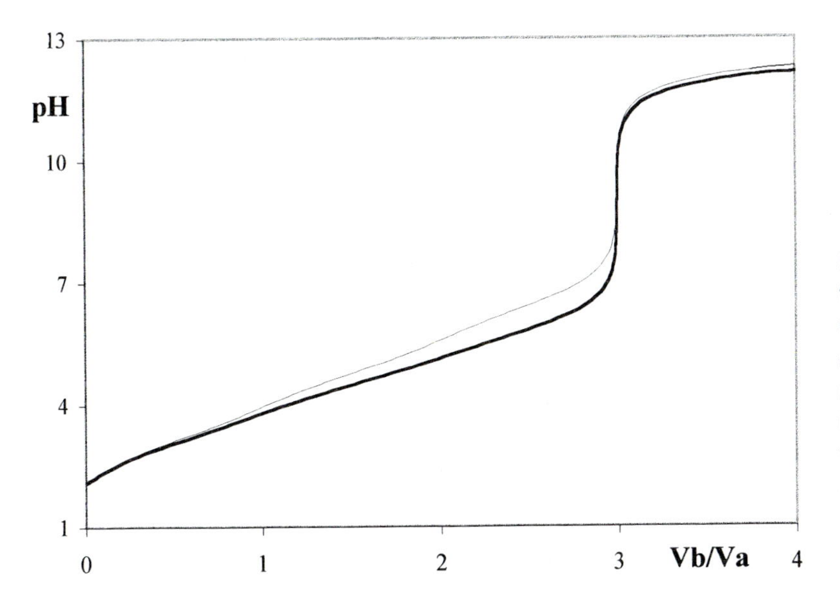

Fig. 6.2 Titration curves for the titration of 0.1 M citric acid with 0.1 M NaOH without (thin line) and with (heavy line) activity correction. All constants used are as in Fig. 6.1. The ionic strength I varies throughout the titration, and is computed by iteration.

varied from 0.0011 M at the beginning of the titration to 0.49 M at the third equivalence point, and to 0.52 M at the end of the titration.

We have used the example of citric acid to illustrate activity effects because these effects are much more pronounced with polyvalent ions, and in the case of citric acid are illustrated rather dramatically because the effect increases gradually during the titration, as the dominant pK shifts from pK_{a1} through pK_{a2} to pK_{a3} to pK_w. Still, even in this case the effect on the titration curve is relatively minor. Although the pH at the equivalence point varies with the activity correction, *the equivalence volume V_e at any equivalence point is not affected at all.*

Similarly, the effect of activity corrections on linear extrapolation methods, such as Gran plots, is usually small, and does not prevent their use. And even when the ionic strength varies during the titration, analysis neglecting activity corrections yields a decent fit, although the numerical values obtained for the pKs will of course be incorrect, as illustrated in Fig. 6.3.

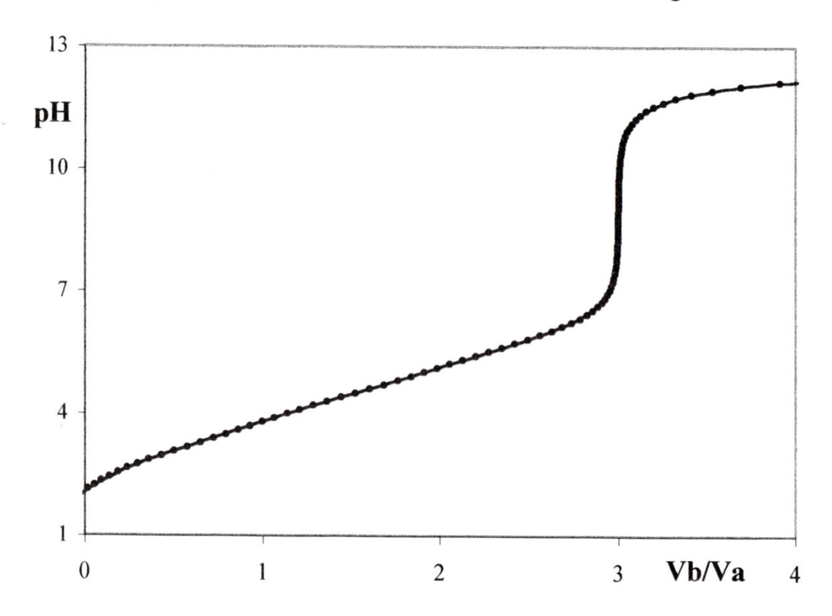

Fig. 6.3 The activity-corrected titration curve calculated in Fig. 6.2 for the titration of 0.1 M citric acid with 0.1 M NaOH (shown in Fig. 6.2 as a heavy line, but here reproduced as *individual data points* as if experimentally measured), and the non-linear least-squares fit of these data to an equation without activity corrections but with adjustable values of pK_w, pK_{a1}, pK_{a2}, pK_{a3}, and C_a (heavy *line*). Values so obtained using Excel Solver and an auxiliary macro to determine the corresponding standard deviations (de Levie 1999b): pK_w = 13.734 ± 0.001, pK_{a1} = 3.027 ± 0.002, pK_{a2} = 4.382 ± 0.002, pK_{a3} = 5.648 ± 0.002, and C_a = 0.10005 ± 0.00002 M, with a standard deviation of ± 0.004 for the over-all fit in V_b/V_a.

In short, activity effects can almost always be neglected when the goal is the accurate determination of sample *concentrations*, as long as the pK-values are considered as adjustable within narrow ranges.

Incidentally, this example clearly illustrates the difference between precision (as stated above in terms of standard deviations) and accuracy, since we started with pK_w = 14.00, pK_{a1} = 3.13, pK_{a2} = 4.76, and pK_{a3} = 6.4.

Activity corrections need only be considered when we wish to obtain estimates for the numerical values of equilibrium constants at infinite dilution without doing the actual extrapolations. And in the latter case one should realize that empirical expressions such as eqn (6.13) based on the *average* behavior of a large number of electrolyte solutions may not fit any *particular* solution very well, so that it is best to use measurements at low ionic strength (say, $I \leq 0.1$ M) where activity corrections are most reliable.

7 The measurement of pH, and its interpretation

In this final chapter we will address some of the practical aspects of pH measurements. While measuring the pH is nowadays both easy and common practice, the interpretation of what is being measured is not quite so simple.

7.1 Spectrometric methods

The pH of a solution can be measured either *electrometrically* or *spectrometrically*. Spectrometry provides many different methods, the most common of which is visible or near-ultraviolet absorption spectrometry using indicator dyes, i.e., acid–base systems that exhibit a pH-dependent change in their spectra. Such measurements are usually more cumbersome than electrometric ones, but their interpretation is straightforward, insofar as Beer's law (1852) involves concentrations rather than activities. Colorimetric acid–base indicators formed the basis of the first quantitative acidity scale, essentially based on unit steps in $-\log [H^+]$ (Friedenthal 1904), even before the pH was defined as such by Sørensen (1909).

Many examples of spectrometric pH measurements and pH titrations are shown in Polster and Lachmann's 1989 fine book on *Spectrometric titrations*. The Hammett acidity function (Hammett and Deyrup 1932a, 1932b, Hammett 1933) is entirely based on spectrometric measurements, and the same is true for its basic equivalent (Bowden 1966). Since these functions find their primary applications in non-aqueous solvents, they are not discussed here; for a critical review the interested reader is referred to the original papers, or to their discussion in Bates (1964) *Determination of pH: theory and practice*, pp. 167–171, or in Galster (1991) *pH measurement: fundamentals, methods, applications, instrumentation*, pp. 228–230.

Alternatively one can use, e.g., fluorescence emission, (again: see Polster and Lachmann's *Spectrometric titrations*), although in that case one should be aware that the fluorescent response may reflect the acid–base properties of both the ground state and the excited state (Förster 1950). Yet another method is to use nuclear magnetic resonance (NMR) spectrometry, see, e.g., Jung et al. (1972).

Acid–base indicators have long been immobilized on paper or other relatively inert substrates, as a quick but crude indication of pH. A more recent development is that of optical sensors involving immobilized acid–base indicators. The relative advantages and disadvantages of such optical pH sensors versus electrometric pH measurements have been discussed by Janata (1987).

7.2 Electrometric pH sensors

By far the most common method of measuring the pH is electrometrically, i.e., by measuring the potential difference between two electrodes immersed in the test solution. This method is fast and easy; unfortunately, its interpretation is not quite as simple, and will concern us in most of this chapter.

The earliest electrode responding to the solution pH is the platinum/hydrogen electrode developed by Nernst (1889), and advocated by him as the basis for a uniform voltage scale, the *hydrogen scale*, thereby replacing the rather arbitrary calomel electrode proposed earlier as a standard by his mentor, Ostwald. Nernst proposed as standard the *normal* hydrogen electrode, obtained with platinum in contact with 1 M sulfuric acid in the presence of 1 atmosphere partial H_2 pressure. (One *normal* H_2SO_4 is 1 M H_2SO_4 because concentrated sulfuric acid acts as a monoprotic strong acid.) This electrode can be made, but for a number of reasons never quite displaced the calomel electrode as a practical reference electrode. Working with a stream of hydrogen gas in a laboratory atmosphere is messy as well as potentially dangerous. The electrode is easily poisoned, and must be replatinized fairly regularly; without the catalytic properties of the platinized layer, the electrode attains equilibrium only sluggishly, or not at all. The electrode is also sensitive to the presence of other reducing and oxidizing agents in solution. Moreover, the use of a concentrated acid leads to a large liquid junction potential difference, see section 7.6. Finally, the standard hydrogen electrode is still based on a somewhat arbitrary choice, as different results are obtained when sulfuric acid is replaced by other strong acids of similar proton concentration, such as 1 M HCl.

After the concept of ionic activity had been accepted, the electrode was redefined as pertaining to a solution of 1 M proton activity, and renamed the *standard* hydrogen electrode. This change was supposed to make the electrode less arbitrary, because its potential would no longer depend on the nature of the anion used. Unfortunately, as was realized only too late, the standard hydrogen electrode cannot be made, since the ionic activity coefficients are neither experimentally accessible nor (outside the range of the Debye–Hückel theory) reliably computable.

What can be done is to use the platinum/hydrogen electrode in much more dilute solutions, such as 0.01 M HCl (where the Debye–Hückel theory can be used to calculate the proton activity) in order to calibrate more practical electrodes, such as the calomel and silver/silverchloride electrodes, on the hydrogen scale. The latter reference electrodes are then used in the actual measurements. The calomel electrode has remained the de facto reference electrode. Neither as a *practical* reference electrode nor as a *practical* proton sensor has the platinum/hydrogen electrode lived up to its promise.

The quinhydrone electrode (Biilmann 1921) was a popular electrometric proton sensor in the 1920s and 1930s, until it was replaced by the glass electrode. It is based on the quinone/hydroquinone redox couple; quinhydrone is the poorly soluble and easily recrystalized complex of quinone and hydroquinone. The quinhydrone electrode is sensitive to the presence of reducing agents such as ascorbic acid. Its use is limited to pH values less than 8; at a higher pH, atmospheric oxygen tends to oxidize the hydroquinone.

The glass electrode, developed by Cremer (1906) and Haber and Klemensiewicz (1909), is the most chemically inert of the available proton sensors. Its use was initially hampered by the large internal resistance of glass. Only after the application of vacuum-tube electrometers to pH measurements (Moore 1922, Calhane and Cushing 1923, Treadwell 1925, Linde 1927, Williams and Whitenack 1927, Morton 1928), and especially to pH measurements using glass electrodes (Elder and Wright 1928, Stadie 1929, Ellis and Kiehl 1933 and references given there, Goodhue 1935), did the glass electrode become the standard tool for electrometric pH determinations. Among its advantages are that the glass electrode is easy to use, chemically rather inert, and can readily be used in colored or turbid solutions. Moreover, its response is unaffected by the presence of either reducing or oxidizing agents.

There are many other electrochemical systems that can be used as proton sensors. Zirconium dioxide can be used in a hydroxyl sensor in much the same way as a glass membrane can be used in a proton sensor. Through the autodissociation constant of water, such an electrode can then function as a hydrogen electrode of the second kind. It is even more chemically inert than glass, and can be used at higher temperatures.

Hydroxide-coated metals such as antimony and bismuth can directly respond to hydroxide ions. Antimony electrodes (Uhl and Kestranek 1923, Kinoshita et al. 1986) are robust, useful from pH 1 to pH 10, and can function in fluoride solutions or under mechanically rugged conditions in which glass electrodes cannot be employed. Aged bismuth electrodes can be used up to pH 16 (Einerhand et al. 1989).

Lipophilic tertiary amines, such as tri-*n*-dodecylamine dissolved in a non-polar medium such as *o*-nitrophenyl octyl ether can act as proton carriers; such solutions can then be used as proton-selective membranes. Lipophilic organic carboxylic acids (the so-called *uncouplers* of oxidative phosphorylation) can also be used as such.

However, among these many options, the glass electrode has become the undisputed standard for electrometric pH determinations in aqueous solutions.

7.3 The electrometric pH measurement

The measurement of pH with a pH meter is deceptively simple. Turn the pH meter on, let it come to thermal equilibrium, calibrate the instrument by inserting the combination electrode (or the separate glass and reference electrodes) in two standard buffer solutions bracketing the anticipated pH reading, then insert the electrodes in the solution to be measured, and read the result. Typically, recalibration does not need to be done too often, so that one can make an entire series of measurements once the pH meter and its electrodes have come to thermal equilibrium and have been calibrated.

First some semantics: the word *electrode* is commonly used in different connotations: it can either mean a metal in contact with a solution, or the entire assembly of metal, surrounding solution, and the vessel holding that solution. When we speak of a hydrogen electrode, a glass electrode, or a reference electrode, we use the second meaning, describing the entire device. In a similar vein, a combination electrode is the object containing two metal wires, each in contact with its own surrounding solution. In order to avoid confusion, this primer we will use the term 'electrode' only for entire devices.

The glass electrode consists of a glass membrane with a fully enclosed solution in contact with a metal wire, see Fig. 7.1. Often that wire is silver, coated with silverchloride, and the enclosed aqueous solution contains chloride ions as well as a pH buffer mixture. The thin glass envelope, often blown out to form a sphere, acts as a H^+ and OH^- exchanger, which responds to the hydrogen ion activities of the adjacent solutions. This leads to a potential difference across the glass membrane which depends on the logarithm of the ratio of proton activities on the two sides of the glass. Since the enclosed solution is fixed, the proton activity of the filling solution is constant (as long as the temperature is unchanged), so that any change in the potential difference reflects a change in the hydrogen activity of the outside solution. There is, of course, another interface in the glass electrode, that between the filling solution and the metal in contact with it, say, Ag/AgCl. However, the resulting potential difference is again constant (at constant temperature), because both the composition of the metal and that of the solution are fixed.

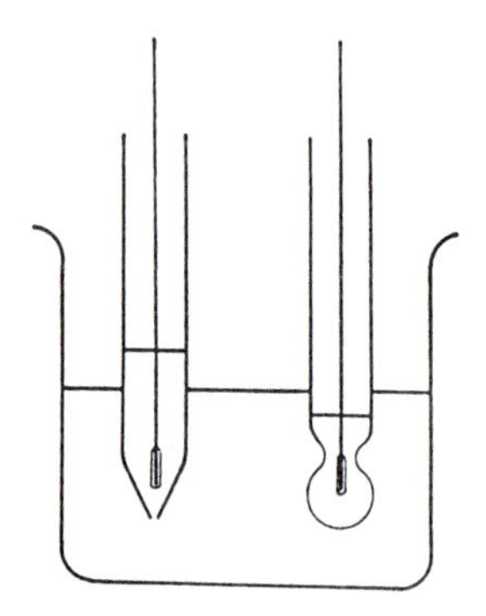

Fig. 7.1 The glass electrode (right) and reference electrode (left) immersed in a test solution. The silver wires are coated with a thin layer (thickness exaggerated in figure) of silverchloride.

The reference electrode (Fig. 7.1) often also contains a silver/silverchloride electrode, typically in contact with a solution saturated in both AgCl and KCl, usually with some excess crystals of KCl to maintain saturation. However, in contrast to the glass electrode, the filling solution of the reference electrode is not completely enclosed, but makes direct, ionic contact with the test solution through a small opening or porous plug, which restricts solution loss as well as contamination of the test solution. From time to time, when the internal solution level gets too low, filling solution must be added to the reference electrode.

The pH meter measures the potential difference between the glass and reference electrode. The meter has two functions, often performed by two successive amplification stages. The glass membrane has a high electrical resistance, typically between 1 and 100 MΩ, and the input resistance of the meter must be much higher than that of the glass electrode, otherwise it will misread the potential difference.

Another way of looking at this is that the measurement must occur with very little current, because even a small current of, say, 1 nA, would generate an intolerably large 'voltage drop' of between 1 and 100 mV across the glass membrane, which would decrease the measured potential difference by the same amount.

The first amplifier must therefore measure the potential difference with negligible input current; for those familiar with operational amplifiers, this can be achieved with a voltage follower, as illustrated in Fig. 7.2.

The second stage provides two functions: it allows for *zero adjustment* (for the *asymmetry* potential difference), and for the translation from potential to pH. The latter involves scaling: at room temperature, the scaling factor is about 59 mV per pH unit, but at 100°C the scale factor is 74 mV/pH. This control is therefore often labeled the *temperature* or *slope* control. The output is then displayed digitally, or on an analog meter.

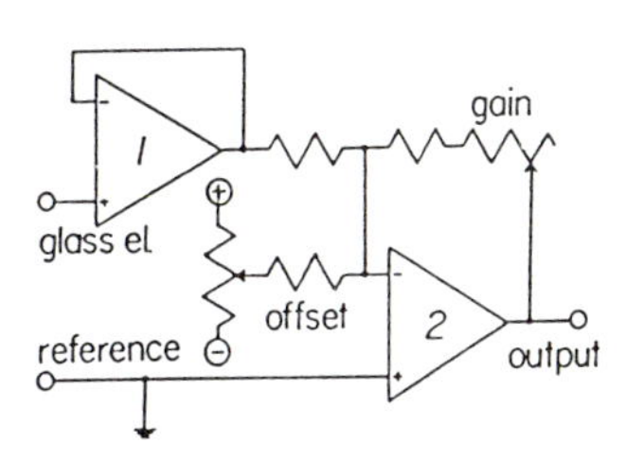

Fig. 7.2 A simple pH meter with a high input impedance unity-gain follower as its first stage, followed by an amplifier with variable gain (often labeled "temperature" or "slope") and offset ("asymmetry").

7.4 What is measured by a pH meter?

In order to understand the use of a pH meter we first need to look at some electrochemical principles and practices. Electrometric measurements determine a potential *difference* between two metal terminals of equal chemical composition, typically made of copper. Like the sound of one hand clapping, the potential of a single terminal or of a single phase is not an *experimentally accessible* quantity. But if you want to study the mechanism of sound production by hand clapping, there is nothing wrong with considering the movement of one hand. Likewise, we will only use such potentials as intermediary results in our calculations, and we will use different symbols to distinguish such immeasurable potentials ϕ and potential differences $\Delta\phi$ from experimentally accessible potential differences E.

According to Kirchhoff's law, any measurable potential difference is made up of the algebraic sum of all the potential differences along the external path from one meter input terminal to the other. First, therefore, we will describe those potential differences. They can be categorized in two classes: potential differences across *bulk phases*, and potential differences across *interfaces*.

Potential differences across bulk phases involve the flow of electrical current; for all practical purposes, they are described by Ohm's law: the potential difference is the product of the resistance of that bulk phase, and the net flow of electrical current through it. In pH measurements with a glass electrode, the net flow of current is minimized by the high internal resistance of the glass membrane, and by the even higher input resistance of the pH meter. Consequently, potential differences across bulk phases are kept at a negligibly low level, and need not be considered here. The potential ϕ in each bulk phase can therefore be considered as constant throughout that phase.

The second category contains all *interfacial* potential differences. These can derive from a number of interfacial processes, such as interfacial charge transfer, interfacial accumulation and depletion (adsorption, ion exchange) of charged particles, or simply by the interfacial orientation of dipolar molecules at phase boundaries. The latter give rise to the so-called *dipole* potential differences. For the present discussion, where the solvent is water throughout the measuring chain, we need not consider dipolar effects.

In section 7.3 and 7.4 we will consider the two dominant types of *interfacial* potential differences encountered in pH measurements, which we will then combine to *measurable* potential differences in section 7.5. Only then can we answer the question posed in the heading of the present section.

7.5 Equilibrium interfacial potential differences

By proper choice and/or design of the interfaces, the potential differences can often be made to respond to the interfacial activities of specific ions. The relation governing this response is the Nernst equation (Nernst 1889), which we already encountered in section 5.5.

The Nernst equation applies when an interface is permeated by only one kind of charge carrier. The contact potential between two dissimilar metals is of that nature: in the interface between, say, copper and silver, electrons are the only mobile charge carriers, and their redistribution at the interface gives rise to the *contact* potential difference exploited in a thermocouple. In this case, the concentrations of the free charge carriers (the conduction electrons) are so high that the interfacial charge transfer is inconsequential to these bulk electron concentrations, and can be ignored, so that the contact potential difference is given simply by a temperature-dependent (but not directly measurable) constant, such as $\phi_{Cu} - \phi_{Ag}$ between copper and silver.

A more relevant example in the present context is the interface between a metal and a solution of its cations, such as that between metallic silver and an aqueous solution of silver nitrate. The mobile charge carriers in the metal are electrons and (at least at the interface, where they can be dissolved or deposited) silver ions, while in solution they are silver, nitrate, hydrogen, and hydroxide ions. In this case, Ag^+ is the only charged species common to both phases, and therefore capable of interfacial exchange. We can consider this a redox process,

$$Ag^+_{aq} + e^-_{met} \rightleftharpoons Ag_{met} \tag{7.1}$$

where aqueous silver ions are reduced to zero-valent silver metal, or silver metal is oxidized to form Ag^+. Alternatively we can describe it as the transfer of monovalent silver ions between water and the 'sea of electrons' of the metal,

$$Ag^+_{aq} \rightleftharpoons Ag^+_{met} \tag{7.2}$$

In either case, the Nernst equation reads

$$\begin{aligned} \phi_{met} - \phi_{aq} &= E^o_{Ag} + 0.059 \log a_{Ag^+_{aq}} \\ &= E^o_{Ag} + 0.059 \log f_{Ag^+} [Ag^+] \end{aligned} \tag{7.3}$$

where $E^o{}_{Ag}$ is the corresponding standard potential, and where we use 0.059 as convenient shorthand for the quantity $RT \ln(10) / F$ which, at 25°C, has the value 0.05916 V. The aqueous silver ion concentration is $[Ag^+]$, while the silver concentration in the metal is constant, and has therefore been included in the standard potential $E^o{}_{Ag}$. The sign of the logarithmic term is most readily found by reasoning, as follows: when the aqueous silver ion concentration is increased, the equilibria (7.1) or (7.2) tend to shift towards the right, so that the metal becomes more positive with respect to the solution, because it loses e^-_{met} or gains Ag^+_{met}. A metal in equilibrium with its own cations yields a so-called potential difference *of the first kind.*

In the presence of solid AgCl we have the additional equilibrium

$$Ag^+_{aq} + Cl^-_{aq} \rightleftharpoons AgCl \tag{7.4}$$

with the associated solubility product

$$f_{Ag^+}[Ag^+_{aq}]\, f_{Cl^-}[Cl^-_{aq}] \rightleftharpoons K_{sp} \tag{7.5}$$

so that eqn(7.3) can be rewritten as

$$\phi_{met} - \phi_{aq} = E^o_{Ag10} + 0.059 \log K_{sp} - 0.059 \log f_{Cl^-}[Cl^-_{aq}]$$

$$= E^o_{Ag/AgCl} - 0.059 \log f_{Cl^-}[Cl^-_{aq}] \tag{7.6}$$

with a potential difference *of the second kind*, responding to anions.

Another mode of interfacial charge transport is often encountered in membranes, i.e., thin layers interposed between two bulk phases. When the membrane is *permselective*, i.e., permeable to only one of the ionic species with which it is in contact, the resulting equilibrium potential again has the form of the Nernst equation. A typical example is the fluoride electrode, which contains a crystal of LaF_3 doped with (i.e., containing a trace of) EuF_2, see Fig. 7.3. The presence of the dopant leads to fluoride vacancies; these can be filled by adjacent fluoride ions, resulting in net fluoride (and vacancy) transport across the membrane. The resulting potential difference between the two solution phases (I and II) of the same solvent, on opposite sides of the membrane, is

$$\phi_{II} - \phi_I = \frac{0.059}{z} \log \frac{a_I}{a_{II}} \tag{7.7}$$

where a_I and a_{II} are the activities of the permeant ions in the solution phases I and II respectively. (Only when I and II have the same solvent does eqn(7.7) lack a standard potential.) The symbol z denotes the valence (including its sign) of the permeant ions. For the fluoride electrode ($z = -1$) we therefore have

$$\phi_{II} - \phi_I = -0.059 \log \frac{a_{F^-_I}}{a_{F^-_{II}}} \tag{7.8}$$

Permselective membranes can be either solid or liquid; an example of the latter is provided by lipid bilayers (or thicker, lipophilic films, held in place by gelation or in a porous matrix) containing valinomycin, a potassium-selective *ion carrier*, for which the Nernst equation yields

$$\phi_{II} - \phi_I = 0.059 \log \frac{a_{K^+_I}}{a_{K^+_{II}}} \tag{7.9}$$

A redox process that involves the metal merely as a donor or acceptor of electrons can give rise to a *redox* or *electron transfer* potential difference. Relevant examples in the context of pH measurement are the redox equilibria

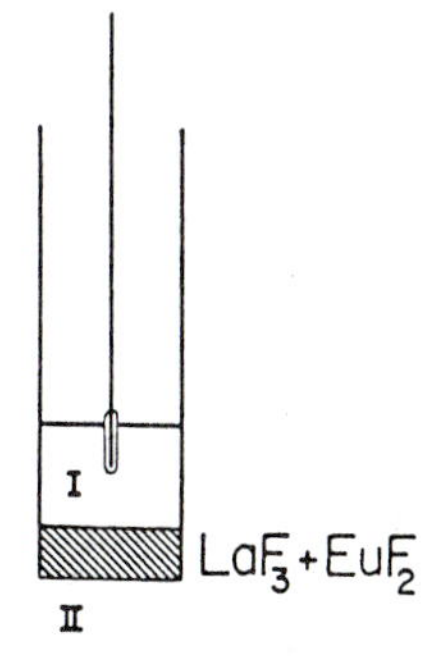

Fig. 7.3 A fluoride electrode consists of crystalline europium-doped LaF_3 separating the internal solution I from the test solution II. The internal solution must contain F^- as well as ions to which the internal metal wire can respond, such as Cl^- when that wire is AgCl-coated silver.

$$2\,H^+ + 2\,e^- \rightleftharpoons H_2 \tag{7.10}$$

and

$$I_3^- + 2\,e^- \rightleftharpoons 3\,I^- \tag{7.11}$$

On normal, smooth platinum, the reaction rates of the proton reduction and hydrogen oxidation are often too slow for the establishment of equilibrium (7.10). For such a *hydrogen electrode* to work, its interface must be made catalytic by platinizing it, i.e., by depositing onto the metal a layer of *platinum black*, which contains highly disordered platinum atoms. The resulting interfacial potential difference then follows the Nernst expression

$$\phi_{met} - \phi_{aq} = E_H^o + \frac{0.059}{2} \log \frac{f_{H^+}[H^+]^2}{f_{H_2}[H_2]} \tag{7.12}$$

which can be simplified when the hydrogen activity in solution is kept constant by maintaining equilibrium between it and hydrogen gas of a fixed partial pressure. The most common partial pressure used is 1 atmosphere (obtained by a steady stream of hydrogen gas bubbling past the platinum), in which case we have the so-called *standard hydrogen electrode*, for which the standard potential is zero by definition, so that

$$\phi_{met} - \phi_{aq} = 0.059 \log f_{H^+}[H^+] \tag{7.13}$$

For the triiodide–iodide electrode, the Nernst equation likewise reads

$$\phi_{met} - \phi_{aq} = E_I^o + \frac{0.059}{2} \log \frac{f_{I_3^-}[I_3^-]}{f_{I^-}^3[I^-]^3} \tag{7.14}$$

An equilibrium interfacial potential difference can also be generated by general or specific ion exchange, and here the glass electrode is a prime example. Glasses are non-crystalline solids in which the molecular arrangements are liquid-like, except that the molecules and ions do not have the mobility of a fluid. Glasses used for pH sensors are usually mixed alkali and earth alkali silicates, where the alkali metal ions are predominantly Na^+ or Li^+, and the earth alkali metal ions mostly Ca^{2+} or Ba^{2+}; their precise compositions are often proprietary. Typical glass compositions (in weight percentages) are 22% Na_2O + 6% CaO + 72% SiO_2 for McInnes glass (Corning 015, Schott 4073) or 25% Li_2O + 8% BaO + 67% SiO_2 for a typical lithium glass such as ENTL 1190. Oxides of trivalent or tetravalent metal ions are sometimes added as network formers, as are heavy metal oxides that prevent devitrification.

In these silicates, the oxygens take up most of the available space, while the metal cations and silicium occupy little more that the interstitial spaces between the oxygens. Upon contact with water, the alkali cations can leach out of the glass, resulting in the formation of a leached glass layer near the interface, typically with a thickness of between 20 and 200 nm (Baucke 1975). Some of the resulting vacancies can then be filled by protons, and this causes the dependence of the resulting *ion exchange* potential difference on the solution pH. The concept that the electrical response of the glass electrode derives from ion exchange goes back to Horovitz (1923); Baucke has refined it to involve specific ions and surface groups (see Baucke 1994).

A glass membrane has two glass/water interfaces, and the resulting potential difference between the aqueous solutions I and II is given by

$$\phi_{II} - \phi_I = E_{asym} + 0.059 \log \frac{a_{H_I^+}}{a_{H_{II}^+}} \qquad (7.15)$$

where the proton activity of the fully sealed filling solution is constant. The *asymmetry* potential difference E_{asym} reflects the differences in the two leached layers at the glass/water interfaces (Bräuer 1941). The value of E_{asym} can be measured in a symmetrical cell, in which solutions I and II have identical compositions, by making contact with these through identical indicator or reference electrodes.

An initial asymmetry may be caused by blowing the glass bulb, when only one side is exposed directly to the flame. Subsequently, after hydration of the glass, the two leached layers may gradually grow further apart in composition, because they lack a mechanism to re-equilibrate, since protons cannot transport through the glass from one leached layer to the other. The inner leached layer, in contact with a fixed solution, may slowly accumulate silver, mercurous or other ions from the filling solution. In the meantime, the outer layer is exposed to different test solutions, and may be subjected to drying out and rehydration. Apart from the asymmetry potential difference E_{asym}, eqn 7.15 is analogous to eqn 7.7.

In summary, equilibrium potentials can be generated by a variety of processes, among them the transfer of metal ions between metal and solution, the transfer of electrons between two components of a soluble redox couple, the selective transfer of ions across a membrane, and ion exchange. The silver/silverchloride and calomel electrodes illustrate electrodes of the second kind based on the metal/metal cation equilibrium. The platinum/hydrogen electrode responds to a redox couple. In a glass electrode, the proton sensitivity derives from specific H^+ and OH^- exchange (Baucke 1994).

7.6 Non-equilibrium interfacial potential differences

The interfacial potential differences described so far depend on the equilibration of only *one* charge carrier (i.e., either on electrons, or on one type of ion) across the interface. Such *equilibrium* potential differences require that the kinetics involved are sufficiently fast, so that equilibrium is indeed established on the time-scale of the experiment. Once such equilibrium has been reached, it is independent of time, and of the geometry of the interface.

When the interface is permeable to more than one type of charge carrier, a different situation exists. For example, in a *liquid junction*, two dissimilar solutions make contact in a way that prevents rapid mixing of the components of the bulk solution phases (typically by making the contact area sufficiently small) but without any selectivity to the ion transport across that interface. Below we will only consider liquid junctions between two *aqueous* solutions; when the solvents on either side of the liquid junction are different, dipole potential differences would also have to be taken into account, and liquid junction potentials can then be quite large (Diggle and Parker 1974).

At a liquid junction between two dissimilar aqueous electrolyte solutions, all types of ions can move freely across the junction, limited only by their individual mobilities and by general electroneutrality constraints, although the latter are somewhat relaxed in interfacial regions that are small compared to the Debye–Hückel parameter $1/\kappa$ encountered in Section 6.2.

Most ions move at comparable speeds, as described by their diffusion coefficients D. According to the Stokes–Einstein relation, the magnitude of the diffusion coefficient is determined by the hydrodynamic (Stokes) radius r of the moving ion and by the viscosity η of the surrounding solution,

$$D = \frac{RT}{6\pi\eta r N_A} \tag{7.16}$$

where N_A is Avogadro's number.

For aqueous solutions, η is mostly constant, and the hydrodynamic radii of many inorganic ions are quite similar, because ions with smaller crystal radii often are more strongly hydrated. (With cations this can even lead to a complete order reversal: hydrated Li^+ has a larger hydrodynamic radius than hydrated Na^+, which in turn is larger than hydrated K^+.) This even applies to solvated protons and hydroxyl ions: their diffusion coefficients, as measured with radio tracer techniques, are quite similar to those of small inorganic cations and anions. The ionic mobility m of a $|z|$-valent ion is related to its diffusion coefficient through

$$m = \frac{|z| D N_A}{RT} = \frac{|z|}{6\pi\eta r} \tag{7.17}$$

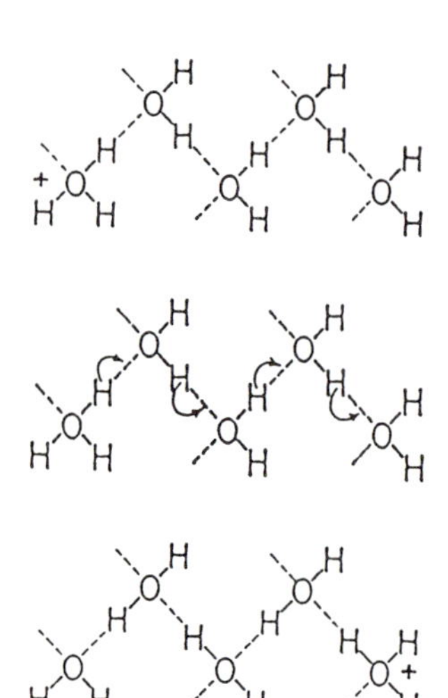

Fig. 7.4 By 'hopping' along its hydrogen bonds, the charge of protons (and hydroxyl ions) can move through water much faster than the actual protons. The three panels show, from top to bottom: an initial state, proton hopping, and the result.

However, the hydrogen-bonded nature of water provides a special, additional mechanism through which protonic charge (though not the individual protons themselves) can move, namely through concerted proton hopping across hydrogen bonds. This mechanism, schematically depicted in Fig. 7.4, involves only small positional shifts of the protons, and can therefore be much faster than diffusion. As a result, when the movement of protonic *charge* rather than of individually labeled (deuterium or tritium) ions is measured, the value is about seven times higher than that of small inorganic cations. The same mechanism is available to hydroxyl ions (which move as proton deficiencies), and similarly results in an enhanced charge transport rate, see Table 7.1.

Whenever two dissimilar aqueous solutions make contact, ions will move across the interfacial region, each kind of ion moving according to the corresponding concentration difference between the two solutions, and with a speed determined by their respective mobilities. If cationic and anionic charges move with the same speed, no net charge is transported across the junction, and no liquid junction potential difference is generated. However, if one kind of ion moves faster than the other, one solution acquires excess positive charge, the other a corresponding excess negative charge. This charge separation leads to a liquid junction potential difference. As that potential difference builds up, it will retard the movement of the faster-moving ions, and accelerate that of its slower-moving counterparts, until both transfer rates are in balance. At that point, further mixing of the components of the two solutions will only involve the transport of neutral salts, and the liquid junction potential difference will have become quasi-stationary. This is not an equilibrium situation: the two solutions will eventually be mixed completely, at which time the liquid junction potential will have disappeared. However, by making the liquid junction sufficiently small (and/or the solution volumes sufficiently large), the time-scale of such complete mixing can be made to be far longer than the measurement time, in which case the liquid junction potential difference can be considered a constant during the measurement.

Table 7.1. Ionic diffusion coefficients D of some monovalent ions at infinite dilution, computed from the aqueous limiting ionic conductances λ at 25°C, as compiled by Robinson and Stokes (1959). The conversion factor is $D = 2.6628 \times 10^{-11}\ \lambda$ when λ is in $cm^2\ \Omega^{-1}\ equiv^{-1}$.

Cation	**diffusion coefficient $/10^{-9}\ m^2\ s^{-1}$**	**anion**	**diffusion coefficient $/10^{-9}\ m^2\ s^{-1}$**
H^+	9.31	OH^-	5.28
Li^+	1.03	F^-	1.48
Na^+	1.33	Cl^-	2.03
K^+	1.69	Br^-	2.08
Rb^+	2.07	I^-	2.05
Cs^+	2.06	NO_3^-	1.90
Ag^+	1.65	ClO_4^-	1.72
Tl^+	1.99	HCO_2^-	1.45
NH_4^+	1.96	$CH_3CO_2^-$	1.09
Bu_4N^+	0.52	$CH_3(CH_3)_2CO_2^-$	0.87

When the two solutions contain i different types of ions, one can derive a general, closed-form expression for the resulting liquid junction potential difference

$$\phi_{II} - \phi_{I} = \frac{RT}{F}\int_{I}^{II}\sum_i \frac{t_i}{z_i}\, d\ln a_i$$

$$= \frac{RT}{F}\int_{I}^{II}\sum_i \frac{t_i}{z_i}\, d\ln c_i + \frac{RT}{F}\int_{I}^{II}\sum_i \frac{t_i}{z_i}\, d\ln f_i \qquad (7.18)$$

where the transport numbers t_i are defined as

$$t_i = \frac{z_i^2 c_i D_i}{\sum_i z_i^2 c_i D_i} \tag{7.19}$$

The transport numbers of the individual ionic species can be determined experimentally. However, the difficulty with eqn (7.18) lies in its integration in the interfacial region, where neither the ionic concentrations c_i nor the corresponding activity coefficients f_i are known in advance.

Below we will briefly describe efforts to evaluate the first integral on the right-hand side of eqn (7.18); evaluating the second term is much more difficult and is usually not even attempted. In order to simplify the discussion, we will assume that all ions have the same valence, i.e., that $z_+ = -z_- = |z|$.

By assuming that the interfacial region remains electroneutral (a clearly invalid assumption in terms of potential differences though, paradoxically, often quite acceptable in terms of ionic concentrations), and that the ionic activity coefficients and diffusion coefficients remain constant throughout the junction, the concentration term on the right-hand side of eqn (7.18) can be integrated when the two solutions only contain one and the same binary electrolyte, and merely differ in its concentration. In that case one obtains

$$\phi_{\mathrm{II}} - \phi_{\mathrm{I}} = \frac{RT}{zF}\,\frac{D_+ - D_-}{D_+ + D_-}\ln\frac{c_{\mathrm{I}}}{c_{\mathrm{II}}} \tag{7.20}$$

where c_{I} and c_{II} refer to the bulk concentrations of the salt in solutions I and II respectively. Even though eqn 7.20 applies to a situation seldom encountered in practice, it illustrates that the liquid junction potential results from the *difference* in the diffusion coefficients of cations and anions. When (in this case of symmetrical z,z-electrolytes) the cations and anions have equal diffusion coefficients, there will be no contribution to the liquid junction potential difference, regardless of the concentration ratio $c_{\mathrm{I}}/c_{\mathrm{II}}$ or, for that matter, of the activity ratio $a_{\mathrm{I}}/a_{\mathrm{II}}$. This situation, in which net cationic and anionic charge moves at equal rates across the junction, is called *equitransference*. In general, equitransference for a single salt requires that $|z_-|D_+ = z_+ D_-$.

In general, integration of the concentration term of eqn 7.18 for liquid junctions involving more than a single binary salt requires *additional assumptions* regarding the nature of the liquid junction; all models to be described below assume constant diffusion coefficients. The additional assumptions can be based either on specific, physically realizable junction geometries, or on mathematical convenience; we will encounter examples of each type.

The first to integrate eqn 7.18 was none other than Max Planck, the discoverer of the quantization of light. (At that time, the concept of an activity coefficient f was not yet known, and was therefore not addressed either.) Planck considered the case in which the junction could be *constrained* to remain within a particular region, such as within a porous ceramic plug or glass frit. The solutions immediately adjacent to the junction would then maintain uniform concentrations, equal to their bulk concentrations. For this situation, Planck (1890) derived his result for solutions that each contain a single 1,1-electrolyte; Planck's treatment can readily be extended to electrolyte mixtures of salts in which all cations have the same valence z_+ and all anions the same valence z_- (Morf 1977). When $z_+ = -z_- = |z|$ one obtains

$$\phi_{\text{II}} - \phi_{\text{I}} = \frac{RT}{zF} \frac{\overline{D}_+ - \overline{D}_-}{\overline{D}_+ + \overline{D}_-} \ln \frac{\sum_i c_{i,\text{I}}}{\sum_i c_{i,\text{II}}} \tag{7.21}$$

While this result *formally* resembles eqn 7.20, it differs from it in that it uses *mean* diffusion coefficients, defined as

$$\overline{D}_+ = \frac{\sum_+ D_+ c_{+,\text{II}} \exp\left[\frac{zF(\phi_{\text{II}} - \phi_{\text{I}})}{RT}\right] - \sum_+ D_+ c_{+,\text{I}}}{\sum_+ c_{+,\text{II}} \exp\left[\frac{zF(\phi_{\text{II}} - \phi_{\text{I}})}{RT}\right] - \sum_+ c_{+,\text{I}}} \tag{7.22}$$

$$\overline{D}_- = \frac{\sum_- D_- c_{-,\text{II}} \exp\left[\frac{zF(\phi_{\text{II}} - \phi_{\text{I}})}{RT}\right] - \sum_+ D_- c_{-,\text{I}}}{\sum_+ c_{-,\text{II}} \exp\left[\frac{zF(\phi_{\text{II}} - \phi_{\text{I}})}{RT}\right] - \sum_+ c_{-,\text{I}}} \tag{7.23}$$

Because these mean diffusion coefficients themselves depend on the liquid junction potential difference, the term $\phi_{\text{II}} - \phi_{\text{I}}$ occurs on both sides eqn 7.21, which therefore is an implicit result that can only be solved by iteration. Nowadays this is no problem, but before the advent of pocket calculators (Morf 1977) or digital computers it could be rather cumbersome to use Planck's equation to evaluate the liquid junction potential difference. Planck's approach was extended to arbitrary ionic valences by Pleijel (1910).

The opposite of a constrained junction is a *free diffusion* junction, in which the boundary is initially sharp, but gradually extends into the adjacent solutions as ions from each solution diffuse into the neighboring solution. MacInnes (1939) described a graphical solution to this problem which, again, can be accelerated greatly by using a computer. A free diffusion junction can be realized in a flowing boundary, and yields a very stable and reproducible liquid junction potential (MacInnes and Yeh 1921, Dohner et al. 1986).

The above two models are based on *physical* assumptions regarding the nature of the liquid junction, and lead to implicit solutions. Below we briefly describe alternative models based on *mathematical* assumptions chosen to facilitate obtaining closed-form solutions. The older of these, due to P. Henderson (1907, 1908), considers the junction as a continuous mixture of the two bulk solutions, i.e., the ionic concentrations at any location inside the junction can be described as

$$c_i = c_{i,\text{I}} + \alpha\,(c_{i,\text{II}} - c_{i,\text{I}}) \tag{7.24}$$

where α varies monotonically throughout the junction from 0 to 1. This assumption allows one to integrate eqn 7.18, and then leads to

$$\phi_{\mathrm{II}} - \phi_{\mathrm{I}} = \frac{RT}{zF} \frac{\sum_i D_i(c_{i,\mathrm{I}} - c_{i,\mathrm{II}})}{\sum_i D_i(c_{i,\mathrm{I}} - c_{i,\mathrm{II}})} \ln \frac{\sum_i D_i c_{i,\mathrm{I}}}{\sum_i D_i c_{i,\mathrm{II}}} \tag{7.25}$$

This result is useful because it illustrates the effect of concentration. If one adds a large excess of an equitransferent salt to one side of the liquid junction, then that salt will have a dominant effect on the sums in the pre-logarithmic terms of eqn 7.25, and thereby make their ratio approach a value of 1. Consequently, we can reduce the magnitude of the liquid junction potential difference by swamping the junction with an equitransferent salt.

Interestingly, in dilute solutions and under the assumptions of Henderson's treatment, the ionic strength I is a linear function of the parameter α used in eqn (7.24), and we can then integrate the rightmost term in eqn (7.18) as long as we remain inside the range of applicability of the Debye–Hückel theory. We then see that, as long as the junction connects two solutions of equal ionic strength I, $d \ln f_i = 0$, in which case the term in the activity coefficients does not contribute to the liquid junction potential difference. Unfortunately, this argument does not apply to the more common situation of a liquid junction between a quite concentrated solution and one that is much more dilute. Still, $\ln f_i$ tends to vary much less than $\ln c_i$, so that the rightmost term in eqn (7.18) is likely to be much smaller than the corresponding term in $d \ln c_i$.

In a model that provides an alternative to that of Henderson, Goldman (1943) assumed that the local potential ϕ is a linear function of distance across the junction (assumed to run from $x = 0$ to $x = 1$), so that

$$\phi = \phi_{\mathrm{I}} + x\,(\phi_{\mathrm{II}} - \phi_{\mathrm{I}}) \tag{7.26}$$

In that case we obtain an very simple result when all ions have the same absolute valence z, viz.

$$\phi_{\mathrm{II}} - \phi_{\mathrm{I}} = \frac{RT}{zF} \ln \frac{\sum_+ D_+ c_{+,\mathrm{I}} + \sum_- D_- c_{-,\mathrm{II}}}{\sum_+ D_+ c_{+,\mathrm{II}} + \sum_- D_- c_{-,\mathrm{I}}} \tag{7.27}$$

Again, by adding an excess of an equitransferent salt to one side, both the numerator and denominator in eqn 7.27 increase by the same amount, thereby making the ratio tend more to 1.

All the above models are similar in that they ascribe the liquid junction potential difference to differences in ionic mobilities. They only differ in the assumptions made about the junction: whether its extent is limited or variable, or whether one can assume a given profile for either concentration or potential. Based on these minor differences, they lead to slightly different numerical predictions. Because these differences are often relatively small (Morf 1977), one often uses the Henderson equation, which is rather easy to evaluate. Table 7.2 lists several such predicted numerical results. Experimentally, the geometry of the liquid junction also matters, especially in terms of stability.

Table 7.2 illustrates that the liquid junction potential differences so estimated can be quite significant, especially in the presence of strong acids, given that one pH unit at room temperature corresponds to about 59 mV. It also shows that the computed liquid junction potential difference becomes smaller in the presence of a high concentration of an approximately equitransferent salt such as KCl. Unfortunately, there is a catch: while the use of a concentrated near-equitransferent salt reduces the magnitude of the concentration term in eqn (7.18), it makes it well-nigh impossible to evaluate the activity term in the same equation, because it brings us far beyond the range of validity of the Debye–Hückel theory.

Table 7.2. Comparison between the predictions of the Planck and Henderson models of liquid junction potential differences, in mV, for three test solutions: HCl, NaCl, and NaOH. Data as reported by Morf (1977), calculated for the potential of a test solution *x* minus that of a salt bridge *sb* containing an essentially equitransferent 4:1 mixture of KCl and KNO_3. As long as the ionic strength I_{sb} in the salt bridge is much higher than that of the test solution, I_x, both models predict quite similar and small values for the liquid junction potential difference.

ionic strength	$\phi_{HCl} - \phi_{sb}$		$\phi_{NaCl} - \phi_{sb}$		$\phi_{NaOH} - \phi_{sb}$	
ratio I_{sb}/I_x	Planck	Henderson	Planck	Henderson	Planck	Henderson
10^4	−0.04	−0.04	0.00	0.00	0.02	0.02
10^3	−0.32	−0.28	0.03	0.03	0.17	0.16
10^2	−2.07	−1.73	0.20	0.20	1.11	1.02
10	−9.40	−8.31	1.11	1.14	5.66	5.27
1	−26.73	−26.77	4.60	4.60	19.35	18.85
10^{-1}	−52.84	−57.58	12.45	12.11	43.54	44.33
10^{-2}	−84.32	−94.06	23.13	22.45	73.24	76.42
10^{-3}	−118.81	−131.95	34.52	33.72	105.24	110.35

7.7 Measurable potential differences

Over most of the aqueous pH range, a glass electrode responds exclusively to the activity of protons. Likewise, in the absence of interfering ions (such as Br^-, I^-, N_3^-, SCN^-, or CN^-), the silver/silverchloride electrode responds to the chloride activity. As a result, the combination of a glass electrode and a silver/silverchloride electrode immersed in a solution of HCl will yield a potential difference that reflects the product of the activities of hydrogen and chloride ions, i.e., the activity a_{HCl} of HCl. For the cell sketched in Fig. 7.5, we have

$$E = \phi_{VI} - \phi_I$$
$$= (\phi_{VI} - \phi_V) + (\phi_V - \phi_{IV}) + (\phi_{IV} - \phi_{III}) + (\phi_{III} - \phi_{II}) + (\phi_{II} - \phi_I)$$
$$= \phi_{Cu} - \phi_{Ag} + E^o{}_{Ag/AgCl} - 0.059 \log a_{Cl,IV}$$

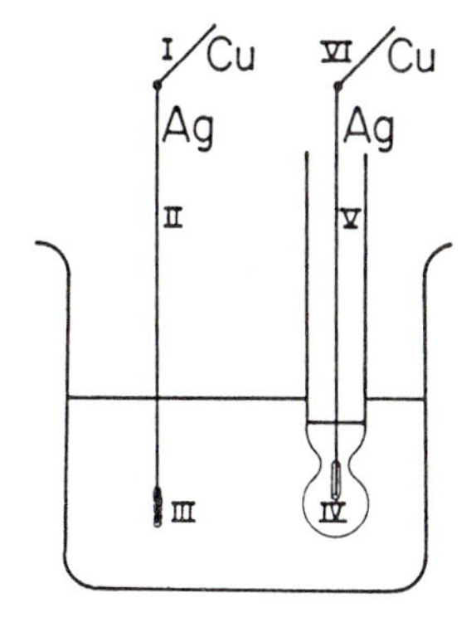

Fig. 7.5 An electrochemical cell without liquid junction can be used to measure *salt* activities. The copper wires connect to silver wires (coated with AgCl) immersed in solutions III and IV; the latter are separated by a thin glass membrane. The silver wires respond to Cl^-, the glass membrane to H^+.

$$+ E_{asym} + 0.059 \log (a_{H,III}/a_{H,IV})$$

$$- E^{o}_{Ag/AgCl} + 0.059 \log a_{Cl,III} - \phi_{Cu} + \phi_{Ag}$$

$$= E_{asym} + 0.059 \log (a_{Cl,IV}/a_{Cl,III}) - 0.059 \log (a_{H,III}/a_{H,IV})$$

$$= E_{asym} + 0.059 \log (a_{H,IV}a_{Cl,IV}/a_{H,III}\, a_{Cl,III})$$

$$= E_{asym} + 0.059 \log (a_{HCl,IV}/a_{HCl,III}) \quad (7.28)$$

Consequently, the activity ratio $a_{HCl,IV}/a_{HCl,III}$ can be determined when the asymmetry potential is measured separately (e.g., with identical solutions in III and IV, so that $a_{HCl,IV} = a_{HCl,III}$).

This is, indeed, how thermodynamic measurements of electrolyte activities are made: find electrodes responding (without interfering, slow kinetics) to the anion and cation of an electrolyte, and measure the resulting potential difference. Make such measurements at various concentrations, and extrapolate the results to infinite dilution, where the activity coefficient is 1.

Unfortunately, this is not what is desired in a pH measurement! In the above example, one electrode responds to the anions, the other to the cations of the test solution. When both act as *indicator electrodes*, the measured potential difference responds to the *electrolyte* activity, in this case to a_{HCl}. That would be fine if we only want to measure pH in solutions provided that these always contain chloride of fixed activity, but that is not what we are after. What we would like to have is an indicator electrode (such as the glass electrode) which responds to *one* kind of ions in solution (such as the hydrogen activity), combined with a *reference electrode* that responds to *none* of the ions in the test solution, but still generates a well-defined interfacial potential difference between the test solution and a metal. Unfortunately, such an ideal reference electrode does not exist. We therefore settle for a close approximation.

We have seen that we can reduce (though not entirely eliminate) the liquid junction potential by swamping that junction with an excess of an approximately equitransferent salt, such as KCl or, better yet, CsCl (Mussini et al. 1990) or RbCl (Buizza et al. 1996); the latter salts are more nearly equitransferent, and can be used at even higher concentrations than KCl. Alternatively, equitransferent salt mixtures can be used.

We can then use a *salt bridge* (Tower 1896) containing such a concentrated solution to link the test solution to a second solution of constant composition, in which we use an appropriate indicator electrode. Or the salt bridge and second solution can be combined in a single *reference* electrode. Examples are Ag/AgCl or Hg/Hg_2Cl_2 in concentrated (often saturated) KCl or CsCl. Of these, the silver/silverchloride electrode is somewhat easier to make. The Ag/AgCl electrode may leak considerable amounts of silver ions into the test solution, because the silver solubility in concentrated chloride solutions is fairly high. The saturated calomel electrode is somewhat bulkier and more difficult to manufacture, and can instead leak mercurous ions. Both reference electrodes have excellent long-term stabilities. Fairly recently,

Ross has introduced another reference electrode, based on the iodide/tri-iodide equilibrium, eqn (7.11), which exhibits a very small temperature dependence.

We would be able to measure the proton activity with a glass electrode (or any other proton-sensitive indicator electrode) and a reference electrode if the liquid junction potential difference could be reduced to zero, or if it could be computed with confidence. It cannot, and neither is there any other known way in which the single ion activity of protons (or of any other kind of ions) can be determined experimentally. Therefore, the pH is no longer defined as the negative logarithm of the hydrogen activity, but either as the (measurable) logarithm of the hydrogen concentration, or as a value believed to be similar to the negative logarithm of the hydrogen activity, but calibrated in terms of standard buffers, see Table 7.3.

buffer	pH
0.05 M potassium tetroxalate	1.679
Saturated potassium hydrogen tartrate*	3.557
0.05 M potassium hydrogen phthalate*	4.008
0.025 M KH_2PO_4 + 0.025 M Na_2HPO_4 *	6.865
0.01 M borax $Na_2B_4O_7.10H_2O$*	9.180
0.025 M $NaHCO_3$ + 0.025 M Na_2CO_3	10.000
saturated $Ca(OH)_2$	12.455

Table 7.3. Four *primary* standard buffers (identified with asterisks) that define pH, and three *secondary* standard buffers, with their pH values at 25°C. They are all at low ionic strength (I < 0.1 M) so that the Debye–Hückel theory can be applied with confidence. Unknown pH values are determined by linear interpolation of the measured pH between the pH values measured for adjacent standard buffers.

A comparison of the response of the glass electrode and a platinum-based hydrogen electrode (MacInnes and Belcher 1931) showed that their glass electrode worked fine between pH 2 and 8. At higher pH-values they found rather significant deviations, which depended on the nature and concentration of the cations; the largest effects were observed in the presence of Na^+. Below pH 2, they observed a somewhat less pronounced dependence on the nature and concentration of the anions.

The strong sodium effect comes about because the glass used turns out to be permeable to Na^+ ions (Schwabe and Dahms 1959), an interesting observation because that same glass exhibits no measurable permeability for H^+! The subsequently developed lithium glasses have much smaller cation-related errors. In general, cation errors of the glass electrode decrease in the order $Na^+ > Li^+ > K^+$; presumably, errors for NH_4^+, Cs^+, and the alkaline earth cations are smaller than those for K^+.

Acid errors are pronounced in hydrochloric acid, are less so in sulfuric acid, and are small in phosphoric acid (Schwabe and Glöckner 1955). Moreover, in HCl and H_2SO_4 these errors are not constant but increase in an approximately linear fashion with contact time. Upon rinsing, such acid errors are retained for quite a while. Glass electrodes should therefore not be presoaked in concentrated HCl solutions, but instead be stored in solutions above pH 2.

7.8 Perspective

In this and the previous chapter we have moved beyond the simple computation of pH, and have considered the uncertainties associated with ionic equilibria: activity effects in Chapter 6, liquid junction potential differences in the present chapter. The theoretical uncertainty in the potential difference across a liquid junction is directly linked to those in single-ion activity coefficients. In fact, there is a vicious circle here. In the words of Harned (1928): "We are thus confronted with the interesting perplexity that it is not possible to compute liquid junction potentials without a knowledge of individual ion activities, and it is not possible to determine individual ion activities without an exact knowledge of liquid junction potentials." In Chapter 6 we showed that activity corrections have only relatively minor consequences for analytical titrations. In the present chapter, we have emphasized the uncertainties in the liquid junction potential difference. How can we reconcile these two?

In the titration of an acid with a base we determine the total amount of replaceable protons. While the shape of the titration curve depends on the activity coefficients, the magnitude of the equivalence volume, V_e, the crucial parameter in an analytical titration, does not. And the liquid junction potential has no effect on V_e either, as long as the junction potential difference is constant, and only a minuscule effect even when it is varies.

On the other hand, when a pH reading is used to calculate the proton concentration, we try to determine the activity of the free protons. In that case, the activity coefficient is directly involved, as is the magnitude of the liquid junction potential difference. Thus, the above two measurements not only determine different quantities, but have quite different precisions: titrations can be very precise; while ionic concentrations or activities computed directly from potentiometric readings seldom are.

We now consider the electroneutrality condition. We have used that condition extensively in Chapters 1 through 5 of this primer, and it is an integral part of the proton condition. Below, we will first look at charge separation in equilibrium and non-equilibrium interfaces, including liquid junctions, and address the paradox mentioned in section 7.6: the assumption of electroneutrality in order to compute a liquid junction potential difference. If electroneutrality were strictly obeyed, there would be no charge separation, hence no potential difference. How can we understand this?

The answer is that electroneutrality is *not* a fundamental law of nature, but a convenient and usually amazingly close approximation in chemistry. Interfacial charge separations leading to equilibrium potential differences typically occur over very small distances: as scanning tunneling microscopy shows, the interface between a metal such as silver and the adjacent solution is very well defined, certainly to well within 1 nm. Imagine that we have two layers of opposite charge density Q, separated by a dielectric medium of permittivity $\varepsilon = \varepsilon_{rel}\,\varepsilon_o$ and thickness d. According to standard electrostatics, this generates a potential difference

$$\Delta\phi = \frac{Q\,d}{\varepsilon} \tag{7.29}$$

In the case of the liquid junction, where the charges are ionic, the charge separation is due to excess ions of one kind over another, and we can therefore write

$$Q = \pm F \sum_i z_i c_i \tag{7.30}$$

We now combine eqns (7.29) and (7.30) to

$$\sum_i z_i c_i = \frac{\varepsilon\,|\Delta\phi|}{F\,d} \tag{7.31}$$

where $\Sigma\, z_i c_i$ is the deviation from electroneutrality, since it would be zero if electroneutrality were obeyed. Now we are ready for our estimate. Since $\varepsilon_0 = 8.9\times10^{-12}$ F m^{-1} and ε_{rel} has a value which usually lies between 1 and 100, we will assume for ε the round value of 10^{-10} F m^{-1}. Likewise, for the Faraday, $F = 96487$ C mol^{-1}, we simply substitute 10^5 C mol^{-1}. We set $|\Delta\phi| = 10$ mV $= 10^{-2}$ V. That leaves us the value of the layer thickness d, for which we will here assume 1 nm $= 10^{-9}$ m. Substituting all of these into eqn (7.31) yields $|\Delta c| = 10^{-8}$ mol m^{-2} or, for a more realistic interfacial area of 1 mm^2 $= 10^{-6}$ m^2, $|\Delta c| = 10^{-14}$ mol mm^{-2}.

Now we can argue about the order-of-magnitude assumptions made, but one thing is clear: 10 mV is a sizable electrical effect (corresponding to more than 0.2 pH units) yet in our example it is caused by an excess concentration of the order of 10 femtomoles. At the usual solution concentrations and volumes, say 1 mL of a 1 mM solution (containing $10^{-3}\times10^{-3} = 10^{-6}$ moles), an excess or deficiency of 10^{-14} moles will not be chemically detectable. This is why electroneutrality is such a good approximation in terms of chemical concentrations, and why we can use it without any qualms in formulating the proton condition.

The only parameter with which you can quibble is the charge separation d, for which we have used 1 nm. For instance, you might argue that the leached layer in a glass electrode is typically of the order of 20 to 200 nm. But even if we increase d by a factor of 10^3, the above argument still holds, because it leads to an excess of 10^{-11} moles in a volume containing 10^{-6} moles. We could detect such a deviation from electroneutrality only if we could measure concentrations to better than 1 in 10^5. For most interfacial potential differences, then, electroneutrality is a very good approximation.

The above argument is not quite applicable to the leached layer, because there the charges are distributed throughout the layer rather than located on specific planes, as in a capacitor, and the area involved is much larger. Similarly we anticipate a distributed charge density in a liquid junction, which can be of much larger dimensions: the porous plug connecting the inner solution of a reference electrode with the test solution can have a length of the order of a mm.

As above, we will consider a liquid junction potential difference of 10 mV over a distance d. Moreover, we will assume that the potential in that region is a monotonic function of distance d, as it would be in all models discussed in section 7.6 We then have an electric field $d\phi/dx$ of the order of magnitude of $\Delta\phi/d$. For a distributed charge we now use the Poisson equation, which we already encountered for spherical symmetry as eqn (6.4), and which in planar geometry reads

$$\frac{d}{dx}\varepsilon\frac{d\phi}{dx} = -\rho = -F\sum_i z_i c_i \tag{7.32}$$

Integration from $x = 0$ to $x = d$ then yields

$$\frac{\varepsilon}{F}\left(\frac{d\phi}{dx}\right)_{x=0} - \frac{\varepsilon}{F}\left(\frac{d\phi}{dx}\right)_{x=d} = \int_0^d \sum_i z_i c_i \, dx = \left\langle \sum_i z_i c_i \right\rangle d \tag{7.33}$$

where $< \Sigma\, z_i c_i >$ is the average deviation from electroneutrality. When we now substitute $\varepsilon = 10^{-10}$ F m^{-1}, $F = 10^5$ C mol^{-1}, and $\Delta\phi / d$ for $d\phi / dx$, we find that $< \Sigma\, z_i c_i >$ is of the order of $\varepsilon\, \Delta\phi \,/\, F\, d^2$ or $10^{-17} / d^2$ mole m^{-3}. For $d = 1$ mm $= 10^{-3}$ m that yields a deviation from electroneutrality of 10^{-11} mole m^{-3} or 10^{-8} mole L^{-1}, much smaller than the usual ionic concentrations in such a junction. Again, we see that minuscule deviations from electroneutrality in terms of ionic concentrations can lead to quite measurable potential differences, and vice versa: we can almost always assume electroneutrality, even when we observe measurable potential differences indicating a charge unbalance. In other words, electrical measurements are enormously more sensitive to small deviations from electroneutrality than are chemical measurements, especially in fairly concentrated electrolyte solutions.

We now return to the question posed in the opening paragraph of this section. Apart from the theoretical uncertainties in the various models for the liquid junction, there is the matter of experimental sensitivity. We have seen in section 6.5 that activity coefficients usually have a relatively minor effect on titration curves, and no effect whatsoever on the equivalent points. This is why we could disregard activity corrections in the first five chapters of this primer. But the situation is quite the reverse when our goal is to determine activity coefficients from electrometric measurements. Salt activity coefficients are typically known to three significant figures. If we aim for a comparable precision of 0.1% in ionic activity measurements, these measurements must be reliable to ± 0.0043 pH units (since log 1.001 = 0.0043) or to ± 0.025 mV. With practical liquid junctions, this is well beyond the present realm of measurement reproducibility, even if there were no questions regarding what theoretical model to use.

The glass electrode is subject to interference at both extremes of the pH range, but at more moderate pH values the major uncertainties in both the pH measurement and its interpretation are usually associated with the reference electrode. The experimental uncertainty derives from the difficulty in making a practical yet reproducible and stable liquid junction; in this respect the free-flowing junction appears to be optimal, but it has not caught on in practice. The theoretical difficulty is more fundamental: the potential difference across the liquid junction cannot be measured directly, and its calculation would require advance knowledge of single-ion activity coefficients, which is unavailable outside the range of applicability of the Debye–Hückel theory.

While pH measurements can easily be read to two (or three) decimal places, such readability is misleading: even very carefully controlled studies of the reproducibility of pH measurements exhibit standard deviations of ± 0.05 pH units (Metcalf 1987). And such an irreproducibility of ± 0.05 pH units or ± 3 mV translates into a corresponding irreproducibility of about $\pm 12\%$ in the proton activity, because $10^{0.05} = 1.12$.

Titrations yield amounts (or concentrations) with more than an order of magnitude smaller standard deviations. But they take more time, consume their sample, and measure something else, viz. the total amount of neutralizable protons in the sample, rather than the activity of the free protons.

Summary

This primer is focused on the computation of the pH, initially simply defined as $-\log[H^+]$, and on pH titrations. For the computation of the pH, double-logarithmic graphs are used to sketch the various concentrations as a function of pH. Such graphs have mostly linear sections with simple (integer) slopes. They can be sketched rather than drawn, because they are used here merely to guide the calculation, rather than to read off specific numerical values.

The computation is based on the proton condition, a single (though not necessarily unique) combination of the mass and charge balance relations, which can usually be written down by inspection. Often, the pH can then be obtained directly by using the graph to identify the dominant terms in the proton condition. In other cases, a quadratic equation may have to be solved. Only in relatively few instances does one need a computer to find the pH; even in those cases, the graphs already yield close estimates.

The discussion of titration curves, in Chapter 3, is also based on the proton condition, and uses a simple master equation to describe all acid–base titrations. This treatment is clearly superior to the more traditional (and highly approximate) piecemeal approach, and readily lends itself to computer fitting of the entire titration curve.

This is followed by short chapter on buffers, and a brief illustration of the applicability of these same approaches to other chemical equilibria.

Chapter 6 discusses activity effects. It is placed *after* the reader has acquired a firm grasp of equilibrium calculations, so that a clear distinction can be made between balance equations (which count concentrations) and equilibrium constants (which deal with activities). The chapter includes an explicit description of how to apply activity corrections (to the level of the Davies equation) to electrometric pH measurements, including acid–base titration curves. Because there is no agreed-upon theoretical model for ionic activity coefficients beyond the reach of the Debye–Hückel theory, ionic activities in most practical, fairly concentrated solutions remain elusive. This is why most tabulations only list ionic equilibrium constants extrapolated to infinite dilution. Fortunately, activity corrections have only a minor effect on titration curves.

The final chapter deals with pH measurements, especially with measurements based on a glass and reference electrode. We encountered equilibrium (thermodynamic) potential differences that are independent of the actual mechanism involved, and non-equilibrium (extra-thermodynamic) potential differences that can be geometry-dependent. In this case, the major source of uncertainty (except in strongly basic or acidic solutions, where it can be considerably larger) derives from both the experimental irreproducibility and the theoretical unpredictability of the non-equilibrium potential difference across the liquid junction. This uncertainty, typically of the order of ± 0.1 pH units, certainly justifies limiting the precision of pH calculations to two decimal places, as done in this primer, where the second decimal place essentially provides a computational guard digit.

Bibliography

S. Arrhenius (1887). Über die Dissociation der in Wasser gelösten Stoffe, *Z. Phys. Chem.* 1: 631.

R. G. Bates (1964). *Determination of pH: theory and practice*, Wiley, New York.

F. G. K. Baucke (1994). Investigation of electrode glass membranes: proposal of a dissociation mechanism for pH-glass electrodes, *J. Non-Cryst. Solids* 19 (1975) 75.

F. G. K. Baucke (1994). The modern understanding of the glass electrode response, *Fresenius J. Anal. Chem.* 349: 582.

A. Beer (1852). Bestimmung der Absorption des rothen Lichts in farbigen Flüssigkeiten, *Ann. Phys. Chem.* 162: 78.

E. Biilmann (1921). Sur l'hydrogénation des qinhydrones, *Ann. Chim.* 15: 109.

N. Bjerrum (1905). Über die Elimination des Diffusionspotentials zwischen zwei verdünnten wässrigen Lösungen durch Einschalten einer konzentrierter Chlorkaliumlosung, *Z. Phys. Chem.* 53: 427.

N. Bjerrum (1914). *Samml. chem. chem.-techn. Vorträge* 21: 1.

K. Bowden (1966). Acidity functions for strongly basic solutions, *Chem. Revs*. 66: 119.

W. Bräuer (1941). Über die Ursache des "asymmetrischen Potentials" von Glaselektroden, *Glastech. Ber.* 19: 268.

J. N. Brönsted (1923). Einige Bemerkungen über den Begriff der Säuren und Basen, *Rec. Trav. Chim.* (4) 42: 718.

C. Buizza, P. R. Mussini, T. Mussini, S. Rondinini (1996) Characterization of aqueous rubidium chloride as an equitransferent ultraconcentrated salt bridge, *J. Appl. Electrochem.* 26: 337.

J. N. Butler (1964). *Ionic equilibrium, a mathematical approach*, Addison-Wesley, Reading.

G. Charlot, R. Gauguin (1951). *Les méthodes d'analyse des réactions en solution*, Masson, Paris, p. 48.

M. Cremer (1906). Über die Ursache der elektromotorischen Eigenschaften der Gewebe, zugleich ein Beitrag zur Lehre von den polyphasischen Elektrolytketten, *Z. Biol.* 46: 562.

C. W. Davies (1938). The extent of dissociation of salts in water VIII: an equation for the mean activity coefficient of an electrolyte in water, and a revision of the dissociation constants of some sulphates, *J. Chem. Soc*. 2093.

C. W. Davies (1962). *Ion association*, Butterworths, London.

P. Debye, E. Hückel (1923). Zur Theorie der Elektrolyten, *Phys. Z.* 24: 185.

J. W. Diggle, A. J. Parker (1974). Liquid junction potentials in electrochemical cells involving a dissimilar solvent junction, *Aust. J. Chem.* 27: 1617.

R. E. Dohner, D. Wegmann, W. E. Morf, W. Simon (1986). Reference electrode with free-flowing free-diffusion liquid junction, *Anal. Chem.* 56: 2585.

R. E. F. Einerhand, W. H. M. Visscher, E. Barendrecht (1989). pH measurement in strong KOH solutions with a bismuth electrode, *Electrochim. Acta* 34: 345.

L. W. Elder Jr., W. H. Wright (1928). pH measurements with the glass electrode and vacuum tube potentiometer, *Proc. Nat'l Acad. Sci USA* 14: 936.

S. B. Ellis, S. J. Kiehl (1933). A practical vacuum-tube circuit for the measurement of electromotive force, *Rev. Sci. Instrum.* 4: 131.

G. M. Fleck (1966). *Equilibria in solution*, Holt Reinhart and Winston, New York.

T. Förster (1950). Die pH-Abhängigkeit der Fluoreszenz von Naphtalinderivaten, *Z. Elektrochem.* 54: 531.

E. C. Franklin (1905). Reactions in liquid ammonia, *J. Am. Chem. Soc.* 27: 820.

H. Friedenthal (1904). Die Bestimmung der Reaktion einer Flüssigkeit mit Hilfe von Indikatoren, *Z. Elektrochem.* 10: 113.

H. Galster (1991). *pH measurement: fundamentals, methods, applications, instrumentation.* VCH, Weinheim.

D. E. Goldman (1943). Potential, impedance and rectification in membranes, *J. Gen. Physiol.* 27: 37.

K. H. Goode (1922). A continuous-reading electrotitration apparatus, *J. Am. Chem. Soc.* 44: 26.

L. D. Goodhue (1935). An electron tube nul-instrument for use with the glass electrode and a description of a rugged type of glass electrode, *Iowa State Coll. J. Sci.* 10: 7-15.

G. Gouy (1909). Sur la constitution de la charge électrique à la surface d'un électrolyte, *Compt. Rend.* 149: 654.

G. Gouy (1910). Sur la constitution de la charge électrique à la surface d'un électrolyte, *J. Phys.* (4) 9: 457.

C. M. Guldberg, P. Waage (1864). Studier over Affiniteten, *Forh. Vidensk. Selsk. Christiana* 35, reprinted in H. Haraldsen, *The law of mass action, a centennary volume 1864–1964*, Oslo (1964) 7.

F. Haber, Z. Klemensiewicz (1909). Über elektrische Phasen-grenzkräfte, *Z. Phys. Chem.* 67: 385.

G. Hägg (1940). *Kemisk reaktionslära*, Geber, Uppsala; German-language edition: *Die theoretischen Grundlagen der analytischen Chemie*, Birkhäuser, Basel (1950).

L. P. Hammett, A. J. Deyrup (1932). A series of simple basic indicators I: The acidity function of mixtures of sulfuric and perchloric acid with water, *J. Am. Chem. Soc.* 54: 2721.

L. P. Hammett, A. J. Deyrup (1932). A series of simple basic indicators II: Some applications to solutions of formic acid, *J. Am. Chem. Soc.* 54: 4239.

L. P. Hammett (1935). Reaction rates and indicator acidities, *Chem. Revs.* 16: 67.

H. S. Harned (1928). The electrochemistry of solutions, in H. S. Taylor, ed., *A treatise on physical chemistry*, Van Nostrand, New York, p. 782.

H. S. Harned, B. B. Owen (1958). *The physical chemistry of electrolyte solutions*, 2nd ed., Reinhold, New York, p. 231.

L. J. Henderson (1908). Concerning the relationship between the strength of acids and their capacity to preserve neutrality, *Am. J. Physiol.* 21: 173.

P. Henderson (1907). Zur Thermodynamik der Flüssigkeitsketten, *Z. Phys. Chem.* 59: 118.

P. Henderson (1908). Zur Thermodynamik der Flüssigkeitsketten, *Z. Phys. Chem.* 63: 325.

D. P. Herman, K. K. Booth, O. J. Parker, G. L. Breneman (1990). The pH of any mixture of monoprotic acids and bases, *J. Chem. Educ.* 67: 501.

J. H. van 't Hoff (1877). Die Grenzebene, ein Beitrag zur Kenntniss der Esterbildung, *Ber.* 10: 669.

K. Horovitz (1923). Der Ionenaustausch am Dielektrikum I: die Elektrodenfunktion der Gläser, *Z. Phys.* 15: 369.

A. Horstmann (1873). Theorie der Dissociation, *Ann. Chem. Pharm.* 170: 192.

J. Janata (1987). Do optical sensors really measure pH?, *Anal. Chem.* 59: 1351.

G. Jung, M. Ottnad, W. Voelter, E. Breitmaier (1972). Circulardichroismus- und ^{13}C-NMR-Untersuchungen zur Dissoziation von Aminosäuren, *Z. Anal. Chem.* 261: 328.

E. Kinoshita, F. Ingman, G. Edwall, S. Thulin, S. Głab (1986). Polycrystalline and monocrystalline antimony, irridium and palladium as electrode material for pH-sensing electrodes, *Talanta* 33: 125.

R. de Levie (1992a). *A spreadsheet workbook for quantitative chemical analysis*, McGraw-Hill, New York.

R. de Levie (1992b). A simple expression for the redox titration curve, *J. Electroanal. Chem.* 323: 347.

R. de Levie (1996). General expressions for acid–base titrations of arbitrary mixtures, *Anal. Chem.* 68: 585.

R. de Levie (1997). *Principles of quantitative chemical analysis*, McGraw–Hill, New York.

R. de Levie (1999a). Redox buffer strength, *J. Chem. Educ.* 76: 574.

R. de Levie (1999b). Estimating parameter precision in nonlinear least squares with Excel's Solver, *J. Chem. Educ.* 76: 000.

R. de Levie (1999c). *How to use Excel in analytical chemistry, and in general scientific data analysis*, Cambridge University Press.

S. Levine, J. E. Jones (1969). Interaction of two parallel colloidal plates in an electrolyte mixture of univalent and divalent ions 1: large separation approximation, *Kolloid Z.* 230: 306.

G. N. Lewis, M. Randall (1921). The activity coefficient of strong electrolytes, *J. Am. Chem. Soc.* 43: 1112.

G. N. Lewis (1923). *Valence and the structure of atoms and molecules*, Chemical Catalog Co.

S. Licht (1985). pH measurement in concentrated alkaline solutions, *Anal. Chem.* 57: 514.

T. S. Light, K. S. Fletcher (1985). Evaluation of the zirconia sensor at 95°C, *Anal. Chim. Acta* 175: 117.

E. Linde (1927). Eine neue potentiometrische Titrieranordnung, *Svensk Kem. Tids.* 39: 285; CA 22: 2293.

D. A. MacInnes, Y. L. Yeh (1921). The potentials at the junctions of monovalent chloride solutions, *J. Am. Chem. Soc.* 43: 2563.

D. A. MacInnes, D. Belcher (1931). Further studies on the glass electrode, *J. Am. Chem. Soc.* 53: 3315.

R. C. Metcalf (1987). Accuracy of Ross pH combination electrodes in dilute sulfuric acid standards, *Analyst* 112: 1573.

W. E. Morf (1977). Calculations of liquid-junction potentials and membrane potentials on the basis of the Planck theory, *Anal. Chem.* 49: 810.

C. Morton (1928). The ionisation of polyhydrion acids, *Trans. Faraday Soc.* 24: 14.

P. R. Mussini, F. D'Andrea, A. Galli, P. Longhi, S. Rondinini (1990). Characterization and use of aqueous caesium chloride as an ultra-concentrated salt bridge, *J. Appl. Electrochem.* 20: 651.

W. Nernst (1889). Die elektromotorische Wirksamkeit der Ionen, *Z. Phys. Chem.* 4: 129.

M. Planck (1890). Über die Potentialdifferenz zwischen zwei verdünnten Lösungen binärer Elektrolyte, *Ann. Phys. Chem.* 40: 561.

J. Polster, H. Lachmann (1989). *Spectrometric titrations*, VCH, Weinheim.

C. G. Pope, F. W. Gowlett (1927). A direct-reading hydrogen-ion meter, *J. Sci. Instrum.* 4: 380.

E. R. Rang (1976). pH-computations in terms of hyperbolic functions, *Comp. Chem.* 1: 91.

J. E. Ricci (1952). *Hydrogen ion concentration, new concepts in a systematic treatment*, Princeton Univ. Press.

K. Schwabe, H. Dahms (1959). Versuche zur Frage der Durchlässigkeit von Glaselektroden für Wasserstoffionen mit Hilfe von Tritium-Markierung, *Monatsber. Akad. Wissensch. Berlin* 1: 279.

K. Schwabe, G. Glöckner (1955). Über das elektromotorische Verhalten von Glaselektroden in stark sauren Lösungen, *Z. Elektrochem.* 59: 504.

F. Seel (1955). *Grundlagen der analytischen Chemie und der Chemie in wässrigen Systemen*, Verlag Chemie, Weinheim.

L. G. Sillén (1959). Graphic representation of equilibrium data, in I. M. Kolthoff, P. J. Elving, E. B. Sandell, *Treatise on analytical chemistry*, part I vol. 1, Wiley, New York.

D. D. van Slyke (1922). On the measurement of buffer values and on the relationship of buffer values to the dissociation constant of the buffer and the concentration and reaction of the buffer solution, *J. Biol. Chem.* 52: 525

S. P. L. Sørenson (1909). Enzymstudien II: Über die Messung und die Bedeutung der Wasserstoffionenkonzentration in enzymatischen Prozessen, *Biochem. Z.* 21: 131.

O. F. Tower (1896). Über Potentialdifferenzen an den Berührungsflächen verdünnter Lösungen, *Z. Phys. Chem.* 20: 189.

W. D. Treadwell (1925). Über die Verwendung der Radioempfängerröhre zu elektrometrischen Titrationen, *Helv. Chim. Acta* 8: 89.

A. Uhl, W. Kestranek (1923). Die elektrometrische Titration von Sauren und Basen mit der Antimon-Indikatorelektrode, *Monatsheft. Chem.* 44: 29.

J. Waser (1967). Acid–base titration and distribution curves, *J. Chem. Educ.* 44: 274.

J. W. Williams, T. A. Whitenack (1927). The application of the electron tube to potentiometric titrations, *J. Phys. Chem.* 31: 519.

C. J. Willis (1981). Another approach to titration curves, *J. Chem. Educ.* 58: 659.

Index